How Schrödinger's Cat
Escaped the Box

How Schrödinger's Cat Escaped the Box

Peter Rowlands
University of Liverpool, UK

World Scientific

NEW JERSEY · LONDON · SINGAPORE · BEIJING · SHANGHAI · HONG KONG · TAIPEI · CHENNAI

Published by

World Scientific Publishing Co. Pte. Ltd.

5 Toh Tuck Link, Singapore 596224

USA office: 27 Warren Street, Suite 401-402, Hackensack, NJ 07601

UK office: 57 Shelton Street, Covent Garden, London WC2H 9HE

British Library Cataloguing-in-Publication Data
A catalogue record for this book is available from the British Library.

ISBN 978-981-4644-61-7
ISBN 978-981-4635-19-6 (pbk)

In-house Editor: Ng Kah Fee

Typeset by Stallion Press
Email: enquiries@stallionpress.com

Printed in Singapore by FuIsland Offset Printing (S) Pte Ltd

Preface

In *How Schrödinger's Cat Escaped the Box*, we are going to attempt
something that many people would say is incredibly difficult, perhaps
impossible. We are going to try to lay bare the nature of physics
at its most fundamental level, the level at which its most profound
mysteries occur — mysteries such as: what are space and time? why
does space have three dimensions? why can't time go backwards?
what is matter? We have puzzled about some of these questions for
millennia, and don't seem to be any closer to the answers than our
remote ancestors were, though we have in the process solved many
fundamental puzzles.

The impression usually given is that such problems involve equations and concepts of fearful mathematical complexity, and that only
those who have spent years of rigorous training in mathematics and
physics have any hope of even beginning to understand them. I don't
think this is true. Nature at this level does not insist on privileged
witnesses; we can all participate. I don't believe that anything in this
book requires *prior knowledge* of physics at any level or mathematics
beyond arithmetic and the simplest algebra. In fact, I believe that
the attentive and systematic reader can, in the process, acquire the
technical knowledge to understand at least some aspects of the fundamental nature of physics, including the quantum mechanics suggested
by the title, at a level far beyond that of popular 'hand waving'.

And yet, though this can be done, it is still difficult to do, and
requires a great deal of mental discipline. No one, trained scien-
tist or otherwise, finds this way of thinking easy. Complexity has
nothing to do with it. The difficulty is not intrinsic to the subject.
There is an immense barrier to be overcome, but it doesn't come

from nature's supposedly complicated ways. It comes *from our own habits of thought*. We have to overcome generations of conditioning which makes us want to see nature in a different way to the one in which it really acts. Centuries of collective effort, in which we have forced ourselves to believe in the results of experiments which often go against the grain, have led us in the right direction, but perhaps we haven't yet gone far enough. Only one method has ever worked though we have always been reluctant to use it. We have to take things to extremes of abstraction at the fundamental level, and never accept a more comfortable compromise. The famous quantum mechanical paradox of Schrödinger's cat is symptomatic of our desire to compromise, to hold on to a view of nature which has some tangible connection with our ordinary world. However, if Schrödinger's cat is ever to escape from its confining box, we have to learn how to escape from ours.

Let us try to outline the problem at the heart of all our understanding of physics, so that we can begin to find a way of tackling it. Physics at a fundamental level is undoubtedly difficult. But this is not because it is complex. Rather, it is because it is *simple*. We have difficulty thinking at a fundamental level because the level of complexity required to produce a thinking being means that the environment in which we have evolved has many *emergent* features which simply do not exist at a more fundamental level. It is natural for us to perceive, for example, the solidity of material bodies — one of the most familiar and reassuring aspects of our everyday world — as a truly fundamental thing, one of the most certain that we have. As Dr Samuel Johnson said when he kicked a stone to contradict the views of an idealist philosopher: 'Thus I refutes Berkeley.' And yet science has shown us, as the result of a long process of systematic observation and theorising, that the solidity of material things is only an illusion. Solidity is an emergent property, not a fundamental one. Reduce the scope of observations from the usual range of millimetres to kilometres by 15 orders of magnitude (or 15 successive powers of ten) and we find that there is no such thing as an extended object. There are only dimensionless points interacting with each other with various forces to give the illusion of rigidity and solidity, and even these cannot be located at particular positions in space.

Considering how much we are conditioned by the complexity of our environment, it is astonishing that we are able to think at all at a fundamental level, and imagine things that we can never observe, which are so very different from the objects of our familiar world. So, how can we do this? How is it that we can do fundamental science at all? The answer is a principle that we are only just beginning to understand in a scientific way, though versions of it have been around for centuries. This is that nature tends to repeat itself at different levels. Of course, it is not the actual structures and qualities that it repeats, *but the abstract patterns that underlie them.*

The question that we have posed is: can we access these patterns and identify them? And, if we can, is it possible to make them accessible to the reader who is not a scientific specialist? I believe that the answer, in both cases, is, most definitely, yes. The patterns are identifiable and they can be made accessible. Even the structures that make quantum mechanics necessary are within the grasp of the intelligent lay reader. Certainly, we need mathematics, but, at this level, it is more of a mathematics of permutation and arrangement, of pattern and symmetry, than one of difficult and complex calculation. To make the argument clearer, and to make the patterning more obvious, we have made extensive use of coloured text and diagrams. However, though prior knowledge is not assumed, this does not mean that we are working only at a superficial level. On the contrary, the aim of this exercise is to try to access some very deep levels of physical thought. The most profound and difficult ideas are often those that at first sight seem the most simple. Consequently, I believe that there will be much in this book for the professional physicist, even at the technical level, and there will certainly be many novel concepts and methodologies.

Quantum mechanics is seemingly at the heart of all our difficulties in understanding the nature of physical reality, but I hope to show that it is the key, rather than the problem. However, the only way to make the breakthrough in understanding is to go to an even more fundamental level which is at present completely unknown territory, and to use mathematical ideas which, though long established, are relatively unfamiliar, even among physicists. To understand where quantum mechanics comes from, we have to transform the way we represent it, and this should have a great deal of relevance to practising physicists, especially as the version required seems to be particularly powerful and transparent to an unprecedented degree. Using this formulation, I believe, we will be able to see exactly why quantum mechanics is there and exactly what it means. The confining box will finally be removed.

Among the many friends and colleagues who have helped everyone's favourite cat to make its great escape, I am particularly grateful to John Cullerne, who worked with me on the nilpotent quantum mechanics in some of its earlier stages, and to Chas McCaw and Mike Houlden who made helpful comments on drafts of the book. I am also extremely grateful to Marion Leibl and Richard Calland for the splendid drawings and diagrams which have gone a long way towards making this work become more understandable. Marion was responsible for the illustration on the cover and for Figures 5, 6, 8 and 11–13, and Richard for Figures 1–4, 9, 10 and 17–20.

Peter Rowlands
Oliver Lodge Laboratory
University of Liverpool
June 2014

Contents

Chapter 1

Introduction

Physics seems to take us as close to elemental nature as we can conceivably imagine, and we have endless examples of its predictive power. Yet, there is still something at its very basis that we don't truly understand. Everywhere, nature seems always to go for the simplest possible options, but physics, as presently understood, is far from simple. We have two fundamental theories, quantum mechanics and general relativity, which are very unlike and in some respects contradictory. Both are complex and highly mathematical and don't seem to tell us much about the simplest and most fundamental things, such as why space is 3-dimensional and why time flow only goes in one direction. Is there any route to finding that simplicity in the chaos of present-day physics? Is there perhaps a 'hidden structure' which will show that all the apparent complications ultimately result from something much simpler?

Quantum mechanics is nearly a century old, and by a long way the most successful physical theory ever devised, yet it is still causing major concerns. Successful or not, it has never been totally liked or accepted except as a 'calculating device'. The main reason is that it has repeatedly denied all our expectations, and produced results which have absolutely no respect for our basic intuitions about the nature of 'reality'. One of the most famous examples of its strange refusal to fit in with what we have always believed to be 'normal' has become iconic. This is the dilemma of Schrödinger's cat, enclosed in a box and finding itself dead and alive at the same time until the box is opened. As its predictive capacity and grip on our lives through modern technology have become ever more powerful, so many have tried to find out why quantum mechanics doesn't behave in the manner expected and have retreated into surrounding it with a mathematical wall beyond which no one of sound mind should venture.

The 'alternative' paradigm of general relativity, with its distortions of space and time, has provided an escape route for some, and has led to even more elaborate real space-time structures as the route to success in physics. But general relativity explains only a few physical phenomena compared to quantum mechanics, and only at a very considerable remove from the main structure of the theory itself. It is more of a 'junior partner' than serious competitor, though of course it came earlier in time. If we want to come to terms with the kind of physics that has been developed most extensively in the last century we need to start with quantum mechanics. Here, we need to ask ourselves, is quantum mechanics really weird, as many commentators suggest, or do we really only think so because we have persisted in certain habits of mind which seem to be hard-wired into our brains? Does our negative view of the subject come from the fact that our expectations have not caught up with our formal structures? Is the supposed violation of 'normality' an opportunity rather than a problem?

It seems to me strange that we have for so long believed that our expectations must be true, for they seem to be based on an instinctive reversion to a world view that is unsustainable as a physical theory. Naïve realism, the view that everything exists just as we observe it, cannot be true if we believe in any form of physics, for physics is founded on *explanation*, and explanation is necessarily different from the thing explained. But the same must apply to what we may call semi-naïve realism, or any physics theory that contains any vestige of naïve realism. This is not because quantum mechanics exists, but because the rationale and internal structure of naïve realism contain logical inconsistencies that cannot be overcome. Naïve realism, in the vestigial form that we find in physics, includes a 'natural' combinations of things that cannot possibly be true at the same time — for example, tangible objects and abstract concepts. We can't reach a fundamental level by thinking of some ideas we use as abstract and others as 'real'. It's a type mismatch. It's like believing in the reality of a partial cartoon or a combination of natural and supernatural.

This is what may be called the 'storybook' tradition. The 'storybook' element comes with the desire to retain a hold on 'reality' as something 'tangible', or the desire to make even *abstract* concepts tangible. It had its origin in religion, both prehistoric and early historic, which provided the first way of abstracting from the completely naïve view of the world as being exactly as we see it, though medieval religion later provided the incentive to developing the mathematical treatment of space and time as we know it today. Some level of abstraction seems hard-wired into the human brain, and it would be impossible to say when the abstract concepts of space and time first appeared, but they were used in stories long before they were used in science. Eventually, science, developing from the initial religious impulse, gave us inanimate material objects acting out parts in space and time, and a science of these developed over several centuries, with a great deal of success in explaining the world.

This paradigm was a necessary stage in the development of human knowledge, as abstracted notions slowly emerged from empirical data, and it is still extremely useful, but quantum mechanics seems to be telling us that this stage is now at an end. Because the picture has been successful for such a long period, our expectations have been that it would continue in this vein as we penetrated deeper into nature, but quantum mechanics seems to be telling us that these expectations were wrong. And if we think about the requirements for a fundamental theory — no disparity between elements, no things accepted wholesale without explanation (however familiar) — then such expectations *cannot be valid*. Our 'storybook' picture must fail at the fundamental level.

If we look at it with unprejudiced eyes, our pre-quantum picture of the world is unsustainable. It can't be what nature is like in its fundamentals. We would have been forced to abandon it sooner or later. Quantum mechanics seems to be telling us something, almost screaming at us, 'This is what nature is like'. And there is no escape — this is the idea that incorporates most of physics, and with a success unparalleled by any previous theory. Experiments have shown that it always gives the correct answers and alternatives do not. We should no longer be complaining about its 'strangeness' but asking what is it telling us. Is there a direction in which it is telling us to go? If we don't ask this question we are missing an opportunity — the only opportunity to extend our knowledge to the next level. As long as we don't ask the question 'Why is it right?' but insist on asking the question 'Why is it wrong?', we will face an impasse on any future development.

Quantum mechanics long ago gave us our chance. It created pathways, a new landscape of ideas, beyond any we could have imagined. The message clearly hasn't got through, or we would otherwise be celebrating it as a liberating force, and working at the next stage in development. There are yet more ideas to be uncovered. To do this we have to take it out of the black box we have put it in, to tear down the protective wall to find out what's inside, and expect to be surprised. We have to look at the special things it tells us about the way to go forward. As long as we don't understand quantum mechanics, we'll be chasing chimaeras. We'll be unable, among other things, to understand why, as Galileo said, the 'book of nature' is written in the language of mathematics. We have actually almost intentionally made the mathematics opaque to avoid confronting the issues that quantum mechanics presents. We are using equations that cannot yield more fundamental answers because they are approximations structured in an opaque mathematics. We have treated the theory as a kind of physics engineering.

In fact, if we examine quantum mechanics from a more fundamental point of view, we find that it is not strange at all. It is a kind of ultimate in abstraction based on symmetries which are ready to be discovered, and the symmetries are the key to explaining why mathematics is effective. Eugene Wigner once wrote of the 'unreasonable effectiveness of mathematics in physics'. There is an equally unreasonable effectiveness of physics in mathematics, which is becoming increasingly apparent. We could, for example, ask why + and – exist as dual options, with nothing in between, something we accept without a moment's thought, but must have a profound origin, like the many dualities in physics. Calculus, also, is a subject that violates the way we normally do mathematics. Despite all attempts to prove that it is a rigorous development, its physical origins remain strongly apparent. It is clear that mathematics and physics are in some way symbiotic at their basis; they grow out of the same root.

If we want a physics that looks more like mathematics and comes closer to it in origin, there is only one way forward — to go in the direction of abstraction. Physics has proceeded by a series of abstractions from empirical observations. It was how we started and we don't know any other way to proceed. Historically, we saw matter and constructed a semi-abstract space in which it had to exist. Gradually we refined our abstractions. Physics has always gained when it has moved in this direction. Quantum mechanics says that it should move even further in that direction, and that is the only possibility we should consider. But it is not sufficient simply to go for a mathematical form. It has to be a particular type of mathematics that makes its physical origins explicit. The mathematics has almost to 'grow' out of the physics.

Quantum mechanics has been made to seem difficult and meaningless because the particular form in which it is presented only seems to operate as a calculating device. It has not been shown fully that it can emerge from something else. Contrary to what many people say, the formalisms we use in quantum mechanics are crucial to our understanding. They are not all equally good for our development. The criterion of their usefulness does not depend only on making correct calculations. We must avoid privileging formalisms which we have adopted only as a result of historical development. We need a route to the inside of the subject as well as outside it.

Current mathematical approaches do not make sense of quantum mechanics and show how it opens up the rest of physics. We need some means of more directly incorporating the abstract property of symmetry, which is ubiquitous in physics at the fundamental level. Symmetry is the only way to connect things which appear to be different in type at the fundamental level. It is not an attempt at unification by brute force, but a matter of finding common origins. It will not be part of my argument that we need a new 'model' for physics, rather a new way of looking at it as we have it now. New physics might be created, but we don't need it to explain what we have.

In fact, I hope to show that physics is an interlocking web of symmetries, and that if we can find the key we can unravel the whole lot. Symmetry emerges from symmetry, duality from duality. At present, physics is an incomplete picture with four forces that behave very differently, though somehow we suspect that they should not, at least in some idealised context. We may well suspect that the key symmetry is a 'broken' or hidden one, or it would already be obvious. Very likely, it is broken not so much because the symmetry has failed but because one symmetry can interfere with another. It is a tricky procedure to separate them out, but it can be done. We look for the outline of the symmetry in the picture and then see how we can sharpen it by understanding the other symmetries involved and how they affect each other. An outcome is that we gain an understanding of many aspects of the most fundamental notions in physics, such as space, time, mass and charge, at a level that might not have been thought possible.

In addition to this, I hope to show that there is a mathematics that naturally emerges with the symmetries that determine physics at the fundamental level. It may be new to many readers, even to some physicists, but it follows very simple rules. It has been around for a long time but has been strangely neglected except by people such as software engineers, who, under commercial pressures, will always go for the most efficient possible processes. Its discovery and its catastrophic loss to science for a century are an amazing story, which is still ongoing.

Using this mathematics reveals a mass of important physical principles and structures which depend, not on calculation, but on abstract structures of symmetry. Though we have to do mathematics, we don't need to do *calculations*. This makes our task much easier. We can concentrate on the concepts, not on the numerical results relevant to future experiments. Existing approaches have privileged calculation over general principles, but many general principles can be established without calculation, and calculations alone have no power to establish general principles. Physicists often use terms such as 'hand-waving' and 'heuristic argument' when they refer to approximate arguments that give the outline of a theory or a calculation, but this is not what we are proposing here. The arguments are fully technical and completely worked out, without approximation, but many of the basic and fundamental ideas can be found through symmetry alone.

The sections are based on exact mathematical arguments aimed at establishing general principles. Readers who want to follow through with calculations on *specifics* can read the extensive account I have already published in *Zero to Infinity*.[1] Such calculations provide the technical detail of the theory, but they are independent of the establishment of general principles, which follows a different process. Prior mathematical knowledge will not be assumed in the reader, and the structures are explained from first principles. Colour-coding is used to show that the mathematics isn't much more difficult than manipulating coloured counters. It may even be the fact that it can be done this way that has led to some of the more mathematically sophisticated practitioners ignoring it. Many people assume that it *can't* be that easy — it can.

Simplicity has often had a difficult time in winning through in physics. The last century saw many cases where simple ideas were rejected when first put forward and struggled for some time afterwards to overcome what seem to be strange prejudices in retrospect. There have been several reasons for this. Nature's idea of simplicity is certainly different from ours — based on cold abstraction rather than comfortable familiarity. But historical examples also show that lack of understanding of simplicity has both psychological and sociological, as well as purely scientific, roots. The psychological problems stem from an erroneous belief that complicated things are 'clever' and that 'clever' answers are better, while the sociological problems, which seem to have been the source of the most bitter conflict, arise from the fact that a great deal of intellectual capital has often gone into complexity, inevitably because complexity comes before simplicity.

Many books have been written on quantum mechanics. They include research monographs and textbooks for professionals and students, as well as popularising discussions on its strangeness and mystery. As a practising physicist, I have read books of all these types and gained much from reading them, but readers will soon find that this is a book of an entirely different kind, though it is aimed at all the constituencies of readers they have appealed to. There will be much that, I think, is new even to the professionals but, at the same time, I hope that a more general readership will find it possible to follow the discussion. Ultimately, I hope that we can show that the apparent dilemmas produced by quantum mechanics lead to extraordinary developments in the understanding of what physics means at the fundamental level. Among other things, we will, I believe, see that Schrödinger's cat has already escaped from its confining box. We need to escape from ours.

Chapter 2

Relativity

2.1 I wouldn't start from here

There is a well-known apocryphal story in which a traveller takes a short cut across country to a rather isolated city, let's say, Cambridge, and, after losing his way, asks one of the locals how to get to his destination, only to get the reply: 'If I were going to Cambridge, I wouldn't start from here'. The same seems to apply to fundamental physics. Of course, physics has been very successful in making sense of the world, and it has had a major impact on the lives of virtually everyone born since the beginning of the last century, but we still don't know the point where it starts; and the place where we are now is certainly not where we would have chosen to begin the quest. The problem is that, since the early part of the twentieth century, we have been faced with two seemingly fundamental, but undeniably contradictory theories, one (general relativity) operating on the very large scale — planets, stars, galaxies — and the other (quantum mechanics) on the very small — atoms, molecules and smaller particles of matter. Everything in physics — and in the whole of nature — happens because of the actions of just four fundamental forces: gravity, electromagnetism (which combines the seemingly separate electric and magnetic forces), and the so-called weak and strong nuclear interactions. While general relativity applies to gravity, it would seem that quantum mechanics applies to everything else. This includes what is called the Standard Model of particle physics, which organises all the known information about the fundamental particles, or smallest units of which matter is constructed (electrons, quarks, *etc.*), and their interactions with each other, in a mathematical theory which makes good agreement with experimental results.

The two main theories disagree with each other on a number of significant points. One of the most crucial is that general relativity absolutely forbids information of any kind to be transmitted faster than the speed of light; particles separated from each other by any distance can't communicate instantaneously any changes in their status. It can only happen over a period of time. Quantum mechanics, however, only makes sense if there is some kind of instantaneous 'correlation' between two particles, separated by any distance and anywhere in the universe. Each particle has to have some 'knowledge' of what all the other particles are doing at any given time. Other problems are that the two theories view even something as fundamental as time in very different and quite incompatible ways, while the basic quantum idea, that energy is exchanged between particles of matter in discrete particle-like packets (or 'quanta') and not as a continuous stream, has never been successfully applied in general relativity or any other theory of gravity.

There is also something of a clash of methodologies. The tightly-controlled combination of quantum mechanics and the Standard Model has been reproduced successfully to the point of tedium in millions of observations. Experimenters have competed with each other to find a chink in the armour with increasingly elaborate experiments, so far without success. General relativity, however, has only been tested in its most superficial aspects; gravity is such a weak force that many of the core conclusions may never be verifiable in a controlled experiment. So, many of the most spectacular predictions, like wormholes, time travel, and even infinite gravitational collapse, while highly publicised, are likely to remain purely hypothetical in the foreseeable future. The main area of application is in cosmology, which again offers only limited opportunities for experiment. We can't rewind the clock to check our conclusions on the origin of the universe, and, in such a field, speculation is likely to run riot. The so-called 'Standard Model' of cosmology can never be as certain as the 'Standard Model' of particle physics, and we have to be cautious about any theory which uses one model to comment on the other.

Physicists often talk about finding a 'unified' theory, by which they mean bringing these two great juggernauts together in a comprehensive synthesis. But it may be that this is entirely the wrong thing to do. A bigger problem in physics than the apparent disunity is the lack of a defined starting point. We can begin mathematics, for example, with numbers learned in childhood but we can't do the same for physics. Even if we managed to find some huge and complicated mathematical superstructure in which both general relativity and quantum mechanics had a place, we would still be baffled by what we mean by space and time, why there are any dimensions at all, and how 'real' matter comes to be distributed in space. These answers can only come from a completely different procedure, a conscious attempt to find the starting point. In a sense, when we find this starting point, we will have reached the ultimate in physics — the equivalent of the number system in mathematics. By contrast, a 'unifying' procedure suggests bringing together the incompatible theories like two tectonic plates impacting with a catastrophic collision and very messy consequences.

In the most publicised recent attempts at unification, we find that the fundamental particles, which are just points in the Standard Model, become extended string- or membrane-like structures, and that the familiar pattern of 3-dimensional space has to be extended to 10 or 11 dimensions, but in a way that we have no means of predicting. There are so many options that the chances of finding the 'correct' one (if any) are vanishingly remote, and we are further away than ever from answering the fundamental questions. Whatever virtues string or membrane theory may have, they won't explain many of the questions that physics originally set out to answer. I most certainly wouldn't start from here!

2.2 How physics is structured

To understand both relativity and quantum mechanics, we first have to define what physics is really about. The point we have reached at the present time tells us that the smallest known units of matter, or *fundamental particles*, interact with each other via fundamental forces causing changes in how they move through space in time, which we can then measure. There are just four forces. Gravity is responsible for all the large scale motions in the universe. It is an attractive force between all particles of matter and is the force that keeps the Earth in orbit round the Sun and the one that makes objects fall to the Earth's surface. The electric or electromagnetic force is the main other one observed in everyday life. It is responsible for all of chemistry and biology, also the cohesion of matter, capillarity, surface tension and friction. Its source is a fundamental quantity known as electric charge, which can be either positive or negative. Magnetism is a special aspect of this force, and is automatically produced by moving charges. For centuries, these were the only known agents of change in nature, but in the late nineteenth century, radioactivity was discovered and two new forces were found deep in the structure of matter. These are called the strong and weak nuclear forces.

The ancient theory that matter was composed of very small atoms was vindicated in the twentieth century, when it was revealed that the bulk of the matter in an atom was contained in a tiny *nucleus* at its centre, composed of positively charged particles called protons and uncharged particles called neutrons. Most of the seemingly-solid atom, in fact, turned out to be empty space, with only a diffuse cloud of negatively charged electrons surrounding the nucleus. The protons and the neutrons in the nucleus were held together by the strong interaction, a force much stronger than any other known. Some atoms, however, were unstable and decayed by radioactivity, in which particles of one kind or another would be emitted from the nucleus. The three main types of radioactivity were discovered early on and they are identified by the emission of three entirely different particles and caused by three entirely different forces. Alpha decay, involving the emission of an alpha particle (a combination of two protons and two neutrons), is a version of the strong interaction, with a rearrangement of the remaining protons and neutrons in the nucleus. Beta decay, involving the emission of an electron, is an example of an entirely new force, the weak nuclear interaction. This is the only force that changes the nature of fundamental particles as well as their arrangement, as a neutron in the nucleus becomes a proton. Despite its name, it is not actually weak at all, as it is a key cause of supernovas, the most cataclysmic events known in nature. Gamma decay, involving the emission of a photon or quantum of pure energy, is a product of the electromagnetic interaction. Unlike alpha and betay decays, gamma decay doesn't change the material struture of the nucleus, only the amount of energy it possesses.

Nothing other than these forces causes anything to happen in nature, and the forces only occur between particles of matter. Something in the particles' structures make them susceptible to particular forces. All particles have mass, a measure of the 'quantity of matter' they contain, expressed in kilograms; particles with electric charges respond to the electric force; particles which respond to the strong and weak forces have some kind of property of a similar nature, which are nowadays often called strong and weak charges, though that usage has been general for less than two decades. The strong charge is also called the 'colour' charge because it comes in three varieties, in analogy with the three primary colours.

Only a limited number of particles are known. Many of the ones originally discovered were later shown to be composites of more elementary units. Thus, protons and neutrons are not really elementary particles; they are each made up of three smaller units, called quarks. We now know that there are six types of quark, arranged in three generations, with apparently repeating properties but with progressively increasing masses. They have been given the names up, down, charm, strange, top and bottom, but these are just labels — they don't signify anything in particular. Quarks, as far as we know, don't exist in a free state. They are held together in groups of three by a version of the strong interaction much stronger than that between protons and neutrons in the nucleus. The proton is now known to be made up of two up quarks and one down, and the neutron of two down and one up. The electron is, to the best of our knowledge, truly elementary, and, just as there are six quarks, so there are six particles of this type; they are called leptons because most of them (but not all) are light — the others are the muon, the tau and three types of neutrino (Table 1).

Table 1. Table of elementary particle.

	Quarks	**Leptons**
First generation	up	electron neutrino
	down	electron
Second generation	charm	muon neutrino
	strange	muon
Third generation	top	tau neutrino
	bottom	tau

The six quarks and the six leptons are the only truly elementary particles that we know of. As a group they are called fermions, after Enrico Fermi, who first worked out the statistics of how they behaved in bulk. There are, however, *antiparticles* to all known particles, that is, particles that have all the properties reversed, except mass (that is, they have opposite signs of charge and directions of spin). For some reason, the universe is made up of matter, rather than anti-matter; but antiparticles are produced by various natural processes, and are observed, for example, in cosmic rays arriving at the Earth's surface from outer space. They have relatively short lifetimes because they annihilate on contact with ordinary matter, producing pure energy as a result. However, it is relatively easy to produce some of them in the laboratory; and antielectrons (which are also called positrons) are used regularly in PET (positron emission tomography) scanners.

It would be consistent with current knowledge to say that the whole of physics and all natural processes are ultimately the result of the twelve known elementary fermions interacting via four fundamental forces. The four interactions, however, though similar in some respects, have startling differences in the way they act. It is one of the main aims of physics to unify these interactions; the Standard Model incorporates three of them (all except gravity) into a single mathematical structure, which makes very accurate predictions, but it doesn't, at the moment, seem to have any more fundamental explanation of the kind we would normally demand in fundamental physics. One of the key elements in the structure is the application of quantum field theory, a version of quantum mechanics featuring the actions of the electric, strong and weak forces. Here, the three interactions

Table 2. The four fundamental forces.

Force	Mediating bosons
gravity	graviton?
electric	photon
weak	W^+, W^- and Z^0
strong	8 gluons

are shown to be mediated by another type of 'particle', called bosons (after J. C. Bose, who worked out their statistical behaviour in a way that separates them from fermions). For the electromagnetic interaction, the boson is the photon; for the strong interaction, there are eight 'gluons'; and for the weak interaction there are three weak bosons, W^+, W^- and Z. The jury is still out on whether there is a 'graviton' to mediate gravity. These are not so much independent particles in the sense that fermions are, but 'quanta of the field', or a part of the interaction mechanism between fermions. In principle, then, we can say that all of nature can be described via the interactions of twelve fermions and twelve antifermions (Table 2).

The only way we can see that anything happens in the world is if something is observed to change, but if everything changed at once we couldn't observe anything because we would have no way of knowing what had happened. Something has to remain fixed to act as a standard. Physics is all about one thing changing while another thing remains fixed. The things that change are called variables, and include space and time. The things that stay the same are called conserved quantities and include energy and electric charge. What makes physics particularly difficult in practice is that its fundamental laws are written down in a way that has to make the variability of the variable quantities absolutely explicit. In fact, we are not really allowed to put these quantities directly into our equations. We are only allowed to talk about their *rates of change* with respect to each other. Typically, we use the rate of change of space (distance or length) with respect to time, which we call velocity (v), but we also measure the rates of change of other quantities with respect to distance or time.

To find the velocity of an object in most physical situations we can't simply divide the distance moved in a particular direction, say

x, by the time taken, say t, as this would be only an average value and in real cases it changes all the time. The process is much more complicated, and requires *calculus*. What we have to do is to write down an expression for the rate of change of distance with time, for which we use the symbol dx/dt. Here, dx is the smallest imaginable change in x; it is an infinitesimal change, so small that it can never be expressed as a number; similarly dt is the smallest imaginable change in t. Something of the form dx/dt is called a differential or, sometimes, a derivative. If we plotted the change of x over time on a graph, dx/dt would be the gradient of the graph at any point (Figure 1), this then gives us the rate of change at any instant, not just the average value over some lengthy period. Sometimes, more than two variables are involved; in these cases we take them one at a time and use a different symbol, ∂, as in $\partial x/\partial t$, but the principle is exactly the same. A good example of a rate of change in ordinary life comes from economics, where the rate of inflation (say I) is the rate at which prices (say P) change (usually increase) over time, and is expressed as dP/dt. Often, as we know from experience, this rate of change is subject to change itself, and we can define the rate of change of a rate of change. So dI/dt is the rate of change of the rate of inflation, either increasing or decreasing. Similarly, in physics, dv/dt is the acceleration, which is the rate of change of velocity with time. In physics, acceleration (a) can represent an increase or decrease in velocity, and so can be positive or negative; or it can represent a change in direction.

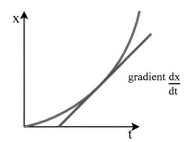

Figure 1. Gradient (velocity) as the rate of change of distance (x) with time (t).

Now, velocity is measured in metres per second, and acceleration in metres per second per second (or metres per second^2). Because a 'rate of' or change with time comes into the expression twice, we call it a second differential or a differential of the second order, and can also write it as d^2x/dt^2. We could also write dI/dt as d^2P/dt^2. Of course, even *this* might change and we could keep increasing the order of the differential until it didn't or until it went in the direction we wanted. Maybe this is what politicians do when they say things like, 'The rate of increase in the rate of inflation is beginning to show signs of slowing down'. For physics, it seems to be the second order differential that is usually important. Squared quantities have a special significance in physics, and we seldom have to consider anything higher. However, even dealing with squared quantities gives us problems.

People often say that physics uses mathematics because it's convenient. It is in fact no such thing. It is very inconvenient. The reason is that the mathematics we use for observation (basically, counting) and the mathematics we use for our equations (calculus) are completely different, and to a large extent incompatible. Because it is based on distinguishing between quantities that are variable and ones that are conserved, physics has to be expressed in *differential equations*, usually with terms of the second order. These are not like ordinary equations with a single or just a few solutions. They contain differentials, which have to be reduced to the ordinary variables by a process called integration, and we have to do this at least twice if there are terms of the second order. The process is nontrivial and doesn't give us a number as the final answer. To find numbers which we can compare with observation, we often have to use special conditions or to make approximations. Differential equations give us general physical laws but they don't themselves give us a series of numbers that we can compare with experiment.

Fortunately, we won't need to do any calculus, to solve any equations, or to calculate any numbers in our discussion, but it is important to see how physics works in the general case. The reason that most physics equations are of the second order is that, in principle, velocity $v = dx/dt$ has no intrinsic significance, but acceleration

$a = d^2x/dt^2$ has. We see this in the laws of motion that Isaac Newton codified for mechanics in the seventeenth century. For these we need two new quantities: momentum (p), defined as mass $\times$ velocity, $p = mv$, where mass is the quantity of matter, measured in kilograms; and force (F), defined as the rate of change of momentum, $F = dp/dt$. Newton's laws then say that a body will continue in a state of rest *or uniform motion in a straight line* unless acted on by an external force. By this we mean that a body doesn't need any force to keep it moving, only to make it accelerate. This is why the Voyager spacecraft is continuing to travel out of the solar system even though its fuel has long since been spent.

This far from intuitively obvious *principle of inertia*, is fundamental to Einstein's special theory of relativity which is designed to make all observers moving with uniform velocity and without acceleration find that the same physics is true as for observers at rest. In principle, there is no way of telling from physics whether you are in uniform motion or at rest. Only force causes the changes that lead to new observable events, and we have long known that there are only four ways of generating force in physics. When we understand all the forces, we will also understand physics because we will know how all types of event occur. Because momentum already contains a differential term (velocity), force, as a differential of momentum, becomes a second order differential, and when the quantity of matter is constant, force can be defined as mass $\times$ acceleration, $F = ma = md^2x/dt^2$.

Significantly, distance (more strictly, displacement or distance in a specific direction), velocity, acceleration, momentum and force are all *vectors* or 3-dimensional quantities. When we apply mathematics to them, we have to take into account their directions as well as magnitudes or sizes, which can be a complicated process. So, while force is the fundamental quantity in physics, and nothing is possible without it, its vector nature makes it awkward to use in calculations, and it is usually more convenient to use other quantities that are related to force rather than force itself. The most important of these is *energy*. When any change is observed due to a force, and the only change that can be observed is a change in spatial position of the force, then an amount of energy is transferred, and the amount of energy is calculated as the force × the distance moved by the force. Energy is a conserved quantity, and the fact that the total amount of energy before and after any interaction is always the same, which is a consequence of Newton's laws, is one of the most fundamental principles in physics. Energy also has the advantage of being a scalar quantity, a quantity with magnitude but no direction, so it doesn't need anything more than ordinary addition. Often the total energy is conserved while changing from one form to another, typically from kinetic (energy of motion) to potential (energy due to the position of a particle in a field or region in which a force might act), or vice versa, sometimes by continual switching between them, as, for example, in the case of a pendulum.

Classical physics could be described as applications of the conservation of energy, or some principle which has the same effect. We define a concept relating mass, space, time, *etc.*, and show that it has some absolute behaviour equivalent to energy conservation. Other conserved quantities include momentum (mass × velocity) and the angular momentum about an axis of rotation (measured as momentum × distance from the axis). Force is structured so that it totals zero within any system that conserves energy, with equal and opposite forces cancelling each other out according to Newton's third law of motion. Other, more subtle principles have the same effect as conservation of energy, for example the principle of least action, which makes the 'action' (a scalar multiple of energy and time or momentum and distance) a minimum value.

Relativity and quantum mechanics do nothing to change this fundamental aspect. Physics equations are still structured to maintain conservation of energy or equivalent (one method of quantum mechanics, the Feynman path integral approach, is based on the principle of least action). Relativity, at least in its restricted or 'special' form, doesn't so much introduce a new physics theory as make corrections to the old one. In our discussion, we will only be interested in those ideas that lead to new knowledge at the fundamental level. Though it is often approached through 'thought' or imagined experiments involving light signals and moving observers, from a fundamental point of view relativity is principally about the mathematical relationship between space and time which emerges from these imagined experiments. In our presentation, we will approach it from this direction, regarding everything else, including the key role of the velocity of light, as a secondary consequence.

2.3 Special relativity

Relativity, as devised by Albert Einstein in the early twentieth century, is not one theory, but two, which go by the names of the 'special' and 'general' theories. Though it is common to give the impression that one theory necessarily leads to the other, and that the general theory, which came later (in 1915), is the inevitable result of the special, which came earlier (in 1905), this is certainly not true. In fact, they are not totally compatible, and modifications have to be made in the special theory if we are to progress to the general. It is also quite possible to believe that one theory is true without in any way accepting the other.

Again, contrary to the impressions given in many popular treatments, the special theory doesn't need any special mathematics, just a version of Pythagoras' theorem, which is the name we give to the method of adding up lengths with different directions in 3-dimensional space (Figure 2). If I travel 3 kilometres due east and then 4 kilometres due north, then this is exactly equivalent to travelling 5 kilometres in a straight line in a direction North 36°52′12" East. I know this because lengths in space at right angles are added using Pythagoras' theorem. If I draw the two lengths I have travelled

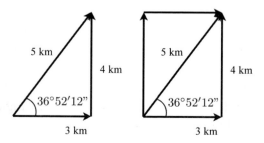

Figure 2. Vector addition: Pythagoras' theorem.

as two sides of a right-angled triangle, then the total distance I have travelled will be the hypotenuse or longest side of the triangle. The sum of the squares of the sides, $3^2 + 4^2 = 25$, will be equal to the square of the hypotenuse, so the length of the hypotenuse will be the square root of 25, or 5. The angle can then be worked out from trigonometry or measured on the diagram. Another way to get the same result is to draw a rectangle based on these two sides, and find the diagonal of the rectangle. If the lengths are not at right angles, we can get the same result by constructing a parallelogram and again finding the diagonal. The process, in physics, is called *vector addition*, and is used for all quantities that have directions in 3-dimensional space (such as length, velocity, momentum, force), as opposed to *scalar* addition, which is used for quantities that have no direction (such as mass, energy, charge).

Space and all other vector quantities are, of course, 3-, not 2-dimensional, and we can easily extend Pythagoras' theorem to a 3-dimensional version by finding the diagonal of a 3-dimensional rectanguloid to calculate the *resultant* or total effect of three *component* vectors at right angles to each other. One way of representing a 2-dimensional vector system would be to draw it on a graph with horizontal and vertical axes, traditionally labelled x and y. Extending this to 3 dimensions would require a third axis (z) imagined as coming out at right angles to the page. In this case, the resultant length (r) would be found by Pythagoras' theorem by adding the squares of the component lengths x, y and z, and taking the square root:

$$r^2 = x^2 + y^2 + z^2.$$

Now, in our original example we could have covered the same distance in the same direction by travelling 4 kilometres east and then 3 kilometres north, or have travelled north first, or even first gone in a direction that was somewhere between north and east. On our graph, we could rotate the axes through some angle, and choose new component values to find the same end result (Figure 3).

This is a standard procedure with all vectors; there is no restriction on the number of ways we can add up component vectors for

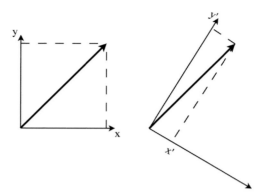

Figure 3. The same vector produced from two different sets of components.

any final desired result. If our new values are x', y' and z', then our new version of Pythagoras becomes

$$r^2 = x'^2 + y'^2 + z'^2.$$

Physicists say that the value of r is *invariant* to the choice of axes as long as we choose the appropriate values of x', y' and z'. We can always find values of components that will make any choice of axes give the same final result, and make

$$r^2 = x^2 + y^2 + z^2 = x'^2 + y'^2 + z'^2.$$

How does this relate to relativity? Well, Einstein realised that, if physical 'information' travelled at the speed of light, then the time delay of information received must be taken into account when we measure a length. After Einstein had worked out all the consequences and developed an elegant set of equations to show how this impacted on contemporary physics, his contemporary and former teacher, Hermann Minkowski found an extremely neat way of representing the idea which was subsequently adopted by Einstein himself. He treated the product of the velocity of light (c) and time (t) as another length (ct), measured in metres, which in a sense it is, and added it as a fourth term to Pythagoras' equation. The only difference is that, to express this as the effect of *delayed* information, the extra term was negative rather than positive, subtracted rather than added.

$$s^2 = x^2 + y^2 + z^2 - c^2t^2 = x'^2 + y'^2 + z'^2 - c^2t'^2.$$

It is notable here, that, although the value of s does not change if we change the coordinate system from x, y, z to x', y', z', the measured value of time does not stay constant but changes from t to t'. Neither the value of overall length nor that of time stays constant when we change the coordinate system, only their *combination*.

Because space is a vector and 3-dimensional, when we add a fourth term we call the result a 4-vector. So 'space-time', as we call it (here represented by s, rather than r), is like a 4-component vector, with the slight difference that one of the terms involves a different sign to the others. Minkowski actually thought that it wasn't just a mathematical construct but something more physically 'real'. Space and time really were one physical quantity: "The views of space and time which I wish to lay before you have sprung from the soil of experimental physics, and therein lies their strength. They are radical. Henceforth space by itself, and time by itself, are doomed to fade away into mere shadows, and only a kind of union of the two will preserve an independent reality."[2]

It is common to rearrange the information about space-time in such a way that

$$c^2t^2 - x^2 - y^2 - z^2 = c^2t'^2 - x'^2 - y'^2 - z'^2.$$

This is called the invariant interval. Now, if x, y, z are all 0, that is if you are travelling with the object, then a particular value of time emerges which we call the proper time. It is given the symbol τ.

$$c^2\tau^2 = c^2t^2 - x^2 - y^2 - z^2 = c^2t'^2 - x'^2 - y'^2 - z'^2,$$

and $c^2\tau^2$ then becomes the invariant interval. The equation implies that the final quantity has axes in a 4-dimensional 'space', which can't be represented on a 3-dimensional diagram.

A number of well-known results follow from this extension of 3-D space to 4-D space-time. If we assume, with Einstein, that the 'velocity of light' is unchanged even if the light comes from a moving source or is detected by a moving observer, we find that moving clocks run slow by a factor $\sqrt{1 - v^2/c^2}$, dependent on the velocity v, and moving objects are contracted in length by the same factor. Also, the mass of a moving object will increase by $1/\sqrt{1 - v^2/c^2}$. Imagine a clock of mass 10 kilograms, 1 metre long, travelling at 90% of the speed of light. It would be slowed down by the factor $\sqrt{1 - 90^2/100^2} = \sqrt{1 - 0.81} = 0.436$. So, compared to 100 seconds on a stationary clock, this one would measure just 43.6 seconds. Its length in the direction of motion would be reduced by the same factor, and so become just 43.6 centimetres, while its mass would be increased by $1/0.436 = 2.29$, so 10 kilograms would become 22.9 kilograms.

It is not just space that transforms from an ordinary vector to a 4-vector, but any vector or 3-dimensional quantity. Particularly significant is *momentum*. This transforms from a 3-dimensional quantity involving components p_x, p_y, p_z, in the three spatial dimensions, into a 4-vector that includes energy as well. The 'invariant interval' this time is called the 'rest mass', or the mass of a system measured when stationary. Here it is the rest mass, rather than energy, that acquires the c^2 term, while momentum acquires a c. The c and c^2 terms are there to give all the quantities the same units — a basic requirement of adding any two things in physics. So

$$m^2 c^4 = E^2 - p_x^2 c^2 - p_y^2 c^2 - p_z^2 c^2,$$

which means that the invariant interval is now $m^2 c^4$.

Though it may be not quite recognisable in this form, this equation is a version of the famous $E = mc^2$. An object that was not moving would have no momentum, so the equation would become $m^2 c^4 = E^2$, which is just a squared form of $E = mc^2$. Even if it is moving, we have a choice of defining m as a fixed quantity (the rest mass) and keeping the momentum as a separate term, or defining a new *relativistic mass*, which *incorporates* the momentum within it, and in this case we return to $E = mc^2$. The justification for doing this is that the relativistic mass then behaves as the mass term did in pre-relativistic physics. Relativistic mass is a particularly useful concept in a particle accelerator, where the particles are accelerated to such high speeds that their 'masses' increase by a large amount, and, when they collide, they disintegrate into higher mass objects.

Physicists generally find equations involving terms with c, c^2 or c^4 too cumbersome. The velocity of light at 3×10^8 or three hundred million metres per second is a man-made number. Babylonian astronomers defined the second and French revolutionaries the metre. We could just have easily defined units so that $c = 1$. In fact we use such units. A light year is a distance, rather than a time measurement; so we say the nearest star (other than the Sun) is about four light years away, meaning that the light will take about four years to reach us. So a speed of 1 light year per year would be a way of defining units so we could make $c = 1$. We could also use light minutes per minute (the Sun is about eight and a half light minutes away) or light seconds per second. In this case, we could write down the energy-momentum equation in the form

$$m^2 = E^2 - p_x^2 - p_y^2 - p_z^2$$

or

$$E^2 - p_x^2 - p_y^2 - p_z^2 - m^2 = 0.$$

Of course, this means that we should then write $E = mc^2$ as $E = m$, which would be strictly correct, though it sounds a lot less impressive!

In these equations, energy substitutes for time and momentum for space. Energy and time, and space and momentum, are known as *conjugate* variables. In a number of cases, particularly in quantum mechanics, these represent different ways of presenting the same information. Space and time give the perspective in terms of the variable quantities, while momentum and energy give it in terms of the conserved ones. Either would be meaningless without the other. They are two aspects of the same overall picture.

Many other consequences follow from adding 'time-like' terms to the three 'space-like' ones in vector quantities. In principle, this is the origin of magnetism as a component of a combined electromagnetic force. Electric fields and force arise from the interactions of electric charges. Magnetic fields and force arise when these charges are observed in motion relative to an observer, so requiring relativistic corrections to the equations for charges that remain static. There is no known independent source for magnetism. At least forty years before special relativity was proposed, James Clerk Maxwell had developed a set of four equations for the electric and magnetic fields which connected them through the velocity of light and predicted the existence of *electromagnetic waves* travelling at that velocity. Special relativity effectively explained the whole structure in terms of the relativistic corrections which needed to be applied to the basic law of interaction between static electric charges.

We have discussed special relativity in relatively abstract terms, avoiding the more usual lengthy discussions about light signals and trains, clocks and observers, using imagined or 'thought' experiments. This is partly because we want to stress the more abstract aspects of physics; but it is also because in many ways it is both the simplest and most direct insight into the real meaning of relativity. It is also, as we will see, the only way of making the idea compatible with quantum mechanics. Ultimately, special relativity works best as an abstract theory, written in mathematical equations. The arguments used to produce this final result are a means to this end, and should not be thought of as ends in themselves. When he first worked out special relativity, Einstein made assumptions which can only be seen as classical approximations to the more fundamental truths provided by relativistic quantum theory. This has often led to widely-stated but quite erroneous conclusions, for example, that special relativity makes fundamental changes in our understanding of the nature of time.

For example, it is often pointed out that events that are simultaneous in one frame of reference (or set of observing conditions) may not necessarily be simultaneous in another. This is true but not in any way profound, for the same thing happens in nonrelativistic physics. For example, I may receive a light signal quite a long time before the sound from the same event reaches me; the well-known delay between a lightning strike and the thunder clap is an obvious example; and we know from geology that older rocks can lie on top of much younger ones. The relativistic examples are nothing more than 'line of sight effects', optical illusions that tell us nothing about time. The 'grandfather paradox' where an apparent reversal of the order of events, taken to an extreme, means that I could, in principle, kill my own grandfather and prevent myself from being born, simply results from using an optical illusion about a time sequence as somehow being connected with the intrinsic meaning of time itself. We could do the same thing without invoking relativity at all.

The relativistic slowing down of moving clocks is often referred to as a 'time dilation', implying an actual alteration in the value of time, but it is simply an artefact of the way we have set up our system of time measurement, *and of the forces we have used to ensure that such measurement is possible.* Special relativity introduces equations involving the differences in time measurements made by observers moving relative to each other with *uniform velocity*; but no experiment could actually ever observe this, and no time measurement could be made in this way. The idea is untenable as a first principle; it is only a convenient heuristic device. All time measurements, even Einstein's preferred method of sending light signals between observers and an observed system, involve force and *acceleration*.

The classic example is the so-called twin or clock paradox. Astronaut A travels from Earth in a spaceship with uniform speed v, while his identical twin brother B remains on Earth. At some point, A turns a round and comes back to meet up with his brother. A's biological clock has run slow by a factor $\sqrt{1 - v^2/c^2}$, and so he should return to see his brother more aged than himself. This is what Einstein believed would happen, and tests with clocks carried round the Earth in aircraft, compared with similar clocks left behind, suggest that this is indeed the case. It doesn't mean that A's *time* is different from B's — the same effect could have been achieved if he'd taken an anti-ageing pill, which would be a different way of applying force and so slowing down his clock. There is no question that force and acceleration are involved — at the moment when the rocket's motors are put into reverse. However, if one follows the logic of the kinematic argument for *uniform* velocity, on which the special theory was founded, then there is no distinction between the stay-at-home twin and the traveller — each would think of themselves as travelling with velocity v with respect to the other, and each would think the other had aged. This is, of course, impossible, and it demonstrates the fact that thought experiments involving uniform velocity tell us nothing about what happens to time. So, while the fact that moving clocks run slow has been demonstrated in many different physical contexts, it has never happened in any situation where acceleration was not involved.

Much is made in many accounts about the distinction between Newton's absolute time and Einstein's relative time, with the implication that a revolutionary change has occurred in our understanding of this concept. The distinction, however, is totally false. *Both* Newton and Einstein use absolute *and* relative time in exactly the same way. Newton writes "Absolute, true, and mathematical time, of itself, and from its own nature, flows equably without relation to anything external, and by another name is called duration: relative, apparent, and common time, is some sensible and external (whether accurate or unequable) measure of duration by the means of motion, which is commonly used instead of true time; such as an hour, a day, a month, a year." But, he also writes: "It may be, that there is no such thing as an equable motion, whereby time may be accurately measured. All motions may be accelerated and retarded, but the flowing of absolute time is not liable to any change. The duration of perseverance of the existence of things remains the same, whether the motions are swift or slow, or none at all: and therefore this duration ought to be distinguished from what are only sensible measures thereof . . . ."[3]

In principle, absolute time, to Newton, is an order of events, not an observable, or what he calls a "sensible measure": "As the order of the parts of time is immutable, so also is the order of the parts of space . . . . All things are placed in time as to order of succession; and in place as to order of situation. . . . in philosophical disquisitions, we ought to abstract from our senses, and consider things themselves, distinct from what are only sensible measures of them.' A complete causal sequence of events, such as one might imagine must exist on a cosmic scale, can never be established by any single observer, because the complete information relating to a physical situation can never be found through measurement. Relative time becomes the order of measured events made without taking into account all the interactions which could possibly affect the measurement.

Newton realises that the *measure of time* is totally determined by applying a system in which force acts, and, if it is dependent on force, then it is dependent on the forces providing an 'equable' measure of motion. If there are frictional forces in watches or perturbations in planetary motions, then the motion will not be equable. In fact, each measure of time will provide a different value because of the different conditions in which each is measured. This is *relative* time. Einstein has exactly the same view of relative time; it will depend on the frame of reference (state of motion) of the observer; and, though frames of reference may be defined in terms of uniform relative motion for the purposes of deriving the relevant equations, in fact this cannot be set up in an experiment. There will always be an acceleration.

For both Newton and Einstein, the true *causal* order of events cannot be violated. To maintain the true order of events, Einstein invokes a 'principle of causality' related to the fact that no information can travel faster than the speed of light, and related to the proper times of the components of the system. This 'principle of causality' is Einstein's equivalent of Newton's absolute time. There is no revolution in our understanding of the fact that one event succeeds another and that this sequence cannot be interfered with however we measure the time.

2.4 Complex numbers

The expression $s^2 = x^2 + y^2 + z^2 - c^2t^2$ is an extension of Pythagoras' theorem to four dimensions, but there is a significant difference. One of the squared terms is negative. If we group all the spatial terms together, *i.e.* do the spatial addition before we add the temporal term, we could represent this as $s^2 = r^2 - c^2t^2$, where $r^2 = x^2 + y^2 + z^2$. Now, if this were an ordinary Pythagorean addition, we could draw a right-angled triangle with adjacent sides r and ct, and hypotenuse s. This would, however, give the wrong answer because it would make s^2 equal to $r^2 + c^2t^2$. The problem is that there is no ordinary number which squares to a negative value: 1 and -1 both square to 1; 3 and -3 both square to 9. Mathematicians realised this was a problem more than four hundred years ago when they were trying to solve quadratic equations, and they quickly came up with an answer. They invented the *imaginary* number i as the square root of -1. Of course, the square root of -1 could equally well be $-i$. The two solutions apply simultaneously, and can't be distinguished from one another, unlike 1 and -1, where we can apply the inequality $1 > -1$ (1 is greater than -1).

The term 'imaginary' should not be taken too literally. All numbers are imaginary in one sense or another. Even the positive integers or natural numbers $1, 2, 3, \ldots$ have no direct correspondence with anything real. There is no circumstance in nature where we can say 2 or 3 objects are exactly identical in all respects, and to use the concept of 1 totally rigorously, we would have to have an impossibly exact definition of the thing to which it applied. We have also long used irrational or algebraic numbers like $\sqrt{2}$, which we can never specify by any ratio of integers, and transcendental numbers like π, which can't even be specified by an equation, without undue concern. In addition, we have a very good diagrammatic representation for imaginary numbers, and for the combination of imaginary numbers with real ones that we call *complex*, for example $2 + 7i$. This is called the Argand diagram and it is a 2-dimensional graph which shows real numbers along the horizontal axis and imaginary numbers along the vertical (Figure 4).

In the example shown, the complex number $2 + 7i$ is represented to scale by a point placed two spaces horizontally from the vertical

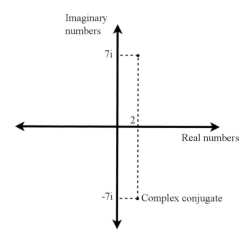

Figure 4. The Argand diagram.

axis and seven spaces vertically from the horizontal axis. The use of the vertical for imaginary numbers is an indication that they can never find a place on the (horizontal) real number line. We can also see it as a kind of 'natural' representation of two dimensions — one that in some way emerges from the mathematics instead of simply being assumed. Mathematics forces us to have two kinds of numbers which are conveniently represented on a 2-dimensional graph.

The Argand diagram doesn't do ordinary Pythagorean addition directly, but we can do it indirectly. If we take a complex number such as $2 + 7i$, then, for anything for which this is a solution, an equally possible solution is the *complex conjugate* $2 - 7i$. On the diagram it is shown as a 'mirror image' in the horizontal axis of the point representing $2 + 7i$. Now, if we multiply a complex number by its complex conjugate, we end up with the sum of the squares of the two numerical values. So $(2 + 7i)(2 - 7i) = 2^2 + 7^2 = 53$. A Pythagorean addition of 2 and 7 would have produced a hypotenuse or resultant length of $\sqrt{53} = 7.28011 \ldots$. We will need to use the complex conjugate when we look at quantum mechanics.

Complex numbers have many uses in physics, but we have introduced them here for a very specific reason. In the case of the Pythagorean addition of space and time, we need to subtract rather than add the squares, so we could represent this by $r^2 + (i^2c^2t^2) = r^2 + (ict)^2$, which is, of course, the same as $r^2 + (cti)^2$. In other words, a convenient representation of time as opposed to space is to give it the status of an imaginary number. It might be possible to argue that this representation is just a 'convenience', but then you have to explain why it is convenient. In fact, there are other indications that the imaginary representation of time is actually physically significant.

2.5 General relativity

The general theory of relativity emerged at a time when people were seriously studying forms of geometry in which Euclid's rules for 3-dimensional space did not apply, especially the so-called fifth axiom which says that parallel lines never meet. In fact, a particular form of non-Euclidean geometry had been well known for centuries.

Spherical geometry applies to the surface of the Earth and is important in navigation. Here the angles of a triangle do not add up to 180° and parallel lines do meet — we call them lines of latitude and they meet at the poles. Einstein's thinking was that if the geometry of real space was not ordinary Euclidean 3-dimensional space but had *curvature*, then an object moving in a region of high curvature might move in such a way as if subjected to a force. A massive object such as the Sun might significantly distort the space around it; or a significant amount of curvature might be considered to be the same as having a massive object there.

At the time when general relativity was first put forward in 1915, the only known forces were gravity and electromagnetism. Radioactivity had been discovered, but the need for two more forces had not yet been established. Gravity had been explained by Newton as an inverse square law force between all objects with the property of mass. That is, the attractive gravitational force between two objects with masses m_1 and m_2 would be given by

$$F = -G\frac{m_1 m_2}{r^2}.$$

Here, we see that the force increases with the two masses involved, but decreases with the distance between them, in fact with its squared value (r^2). By 1915 this law had explained the motions of nearly all the planets, satellites and asteroids in the solar system as being caused by the gravitational attraction of the Sun and all other known astronomical bodies.

The first thing to note is that the force is mutual. An object at the Earth's surface attracts the Earth just as much as the Earth attracts the object. A negative sign is used for an attractive force because it signifies pulling back in, while vector directions point outwards. (This is an example of a useful *mathematical convention*, useful because it provides information about direction as well as size or magnitude.) The quantity G, known as Newton's gravitational constant, arises simply because we have defined lengths in metres, mass in kilograms and time in seconds. Like the velocity of light (c), it is another artefact of the way we have historically defined our units. The equation shows that the force falls off not with distance, but with the squared distance from the source. The reason for this is that the gravitational influence is spherically symmetric. In the eighteenth century the philosopher Immanuel Kant showed that inverse square laws were natural for 3-dimensional space. As we expand in a sphere from the centre, the effect will be diluted by the increasing surface area $(4\pi r^2)$, in the same way as a deep red balloon quickly becomes pale pink when we blow it up.

Let us examine the effect of the force on an object with a mass of 0.1 kilograms or 100 grams at the Earth's surface. The mass of Earth is 5.97×10^{24} kilograms (or 5.97 kilograms multiplied 24 times by 10); the radius of Earth is about 6,370 kilometres or 6.37×10^6 metres; the constant G is 6.67×10^{-11} in corresponding metric units (6.67 divided 11 times by 10). So, the force exerted by the Earth on the object, which is equal and opposite to the force exerted by the object on the Earth:

$$F = -G\frac{m_1 m_2}{r^2} = -6.67 \times 10^{-11} \times \frac{5.97 \times 10^{24} \times 0.1}{6.37 \times 10^6 \times 6.37 \times 10^6}$$
$$= -0.981 \text{ metric units.}$$

The metric unit of force is called the 'newton', so this is 0.981 newtons, and it is this force that we call the 'weight' of the object. The negative sign implies that it is an attractive force, and so directed towards the centre of the Earth. If our object had a mass of 102 grams, it would be subjected to a force of exactly 1 newton, which, by sheer coincidence, is about the weight of an apple from an English orchard. We can now apply mass $\times$ acceleration to find the object's acceleration and the Earth's acceleration. The object's acceleration is $0.981/0.1$ kg $= 9.81$ metres/second^2, which is exactly the acceleration of free fall measured at the Earth's surface. The Earth's acceleration, however, is much smaller, though the Earth is subjected to exactly the same degree of force, attracting it to the object. We can calculate it at $0.981/5.97 \times 10^{24} = 1.64 \times 10^{-25}$ or less than a billion billion billion billionth of a metre/second^2, an absolutely negligible amount.

For every force equation, there is also an energy equation. So we can write the gravitational interaction in terms of the *gravitational potential energy*

$$V = -G\frac{m_1 m_2}{r}$$

rather than the force. For a force that decreases with the square of the distance, the energy decreases with the distance itself. Using this formula, we can calculate that the gravitational potential energy of an object of mass 100 grams at the Earth's surface is about 6,250,000 metric units or joules.

$$V = -G\frac{m_1 m_2}{r} = -6.67 \times 10^{-11} \times \frac{5.97 \times 10^{24} \times 0.1}{6.37 \times 10^6}$$

$$= -6.25 \times 10^6 \,\text{joules}.$$

Just as an attractive force is represented by a negative sign, the energy of such a force is also represented as negative. This signifies the energy *that would have to be supplied* to remove it from its position, which is exactly what we do when we send an object from the Earth's surface into space. The energy that we supply, however, say by using a rocket, will be positive and will be equivalent to providing a *repulsive* force to send the object away from the Earth. Both the force and the potential energy are relatively easily calculated in Newtonian theory using single equations. This is not possible in general relativity.

General relativity sets up an equation in which an array of variables called a *tensor* is used to indicate the degree of curvature to be applied to the combined space-time differentials and is related to another tensor which indicates the energy and momentum components needed to produce this. This is a complicated differential equation or set of equations, and there is no easy substitution of numbers to find answers. Solutions have only been found in the most idealised of conditions. The main, and perhaps only testable, solution is the so-called Schwarzschild solution for the spherical space round a point source. Of course, gravitational sources are not point-like, but large extended objects like the Sun and other stars, but gravity is such a weak force that the approximation is good enough to produce the main effects so far observed: an additional precession or orbital rotation in the orbits of certain astronomical objects; a deflection of light or other radiation by a strong gravitational source, or an equivalent delay in the time for a pulse of radiation to pass by such a source; a shift in the frequencies of radiation emitted or absorbed by a strong source like the Sun.

The deflection of light or other radiation is particularly important, as it was the main experiment used to establish the theory in 1919 by confirming one of its main predictions. Light coming from a distant star which just grazes the edge of the Sun should, according to both Newton and Einstein, be deflected in towards the Sun to some degree. The only chance of seeing this effect is during a total eclipse when the Sun's own light is temporarily extinguished. The star would be seen in a different place to the expected one when compared to stars further away. You would effectively take photographs both during the eclipse and when the Sun was nowhere near to make the comparison. In the Newtonian view, the light particles having a mass (or in a more modern view, the photons having energy and therefore relativistic mass) would be attracted by the force of gravity towards the Sun. In Einstein's view, the Sun would distort the 4-dimensional 'space-time' in its vicinity and the light would follow the lines of curvature (the so-called 'geodesics'). In effect, the light would not 'know' that it wasn't travelling in a straight line (Figure 5).

According to the astronomer Arthur Eddington's interpretation, there would be an effect in Newtonian theory — Newton had hinted at it but had not calculated it — but it would be only half that due to general relativity. The effect is actually very difficult to observe with

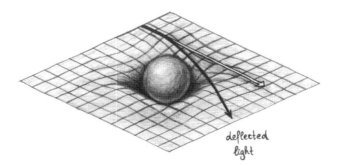

Figure 5. Gravitational deflection of light: distortion of space-time by massive object, like the Sun.

great accuracy using the eclipse method, and has probably never really been observed by this method with the accuracy required. Accuracy is now much easier to achieve using a radar beam, and there is no doubt that the effect is valid, but the measurements led by two expeditions under the control of Eddington claimed vindication of the general relativistic value in a way that led to the immediate acceptance of the entire theory. In the long-term perspective, the more significant issue is not the questionable accuracy of Eddington's results of 1919, but how far any of the tests of general relativity go towards verifying the theory, and what aspects of the theory are actually being tested.

This is important to a generation of physicists who are faced with assessing the claims of two major theories which seem to be in conflict and if it is general relativity that has to be modified, rather than quantum mechanics, we need to know. Ultimately, it seems, according to the most generally accepted picture, that one theory or the other must be abandoned or subjected to modification. This means that both have to be subjected to the same degree of rigorous scrutiny, regardless of historical and cultural impact, and if it is general relativity that has to be modified, we need to know in the greatest possible detail what parts of the theory have stood up to the scrutiny of experimental verification. In particular, we need to know how much of the mathematical structure represented in Einstein's field equations actually supports the ideas which led to their foundation.

However, it may be that the real issue is not about modification, but *interpretation*, and that the real problem is not about physics, but about *language*. A mathematical expression is described by a language of physical interpretation which is not an exact equivalent and which makes claims that are far from direct consequences of the mathematics. Perhaps there is no conflict at all between general relativity and quantum mechanics. They are both essentially abstract generic theories, which have been subjected to various physical interpretations that are not essential to the abstract structures on which they are built. At the abstract level, they are compatible, but the interpretations have chosen different routes to using them to explain observed phenomena. The field equations of general relativity notably actually emerged as a 'best mathematical fit' to a number of conflicting physical requirements — especially those known as the principle of general covariance, Mach's principle, and the principle of equivalence — none of which survived intact in the final theory. The final theory is a mathematical representation of space-time curvature without specific connection to gravity as we know it from observation (which is very much the logic of Einstein's position).

To relate the mathematical structure to anything measurable through gravity, further assumptions have to be made, and restrictions have to be imposed to confine the solutions to special and rather limited cases. Though this is to some extent true of all physics based on differential equations, general relativity, mainly because of the extreme weakness of the gravitational force, is obliged to take it to an extreme. A successful demonstration of the truth of one interpretation of the field equations doesn't necessarily carry all the others with it. General relativity, like the special theory which preceded it, finds its ultmate form as an abstract theory, and this is how it should be integrated into physics. As with the special theory, the abstract form may or may not contradict the 'physical' assumptions which led to its creation; but in any circumstances, it is the abstract form which must prevail, even if this means abandoning some of our previous assumptions. Far from *contradicting* general relativity, this would actually enhance the theory and make it more generally valid. It would mean that the theory would hold up even in the strong gravitational fields provided by such objects as short-period binary pulsars, when alternative interpretations might indicate that deviations should be expected.

There is no doubt that the classic tests based on the Schwarzschild solution — redshift, perihelion precession, light bending and radar time delay — give the correct results. Ultimately, these results come from the idea that gravity produces the equivalent of special relativistic time dilation and length contraction, and that these provide the corrections that appear in the Schwarzschild solution, distinguishing it from a nonrelativistic theory. Much weaker tests are based on the idea that a speed of light connection will produce an equivalent of the magnetic effects in electromagnetic theory and a set of equations similar to those in the standard electromagnetic theory of Maxwell, ultimately leading to 'gravitational waves' of a kind similar to electromagnetic waves. These have not yet been detected directly, but there is indirect evidence of the right kind of energy loss from observations of the behaviour of a binary pulsar. However, even these don't really go beyond the Schwarzschild solution.

In principle, the experimental results are all good evidence, but general relativity as a theory is much bigger than the Schwarschild solution. Newtonian gravity was first verified by Halley's successful prediction of the return of a comet, but that alone would not have been sufficient to vindicate it. It needed systematic application to the whole of the solar system over many decades by Euler, Lagrange and Laplace, and the successful launching of probes in the twentieth century. General relativity makes many claims which have not yet been successfully tested that are far more radical than the consequences of the Schwarzschild solution. So, if we are to include these deeper consequences as an essential element of a more fundamental theory, we have to be prepared to examine in detail how far our experimental results distinguish general relativity from Newtonian gravity and special relativity in particular, and also which *interpretations* of the essentially abstract field equations are really valid in fundamental terms. We should not allow ourselves to assume that successful testing of the Schwarzschild solution is in any way a vindication of all these deeper consequences and interpretations.

The more problematic claims are related to the idea of 'nonlinearity' — that is, the idea that, if gravity causes curvature, then that curvature must increase the gravity, causing further curvature, and so on. If nonlinearity is true, then general relativity is not a truly fundamental theory, but only a best-fit approximation to be superseded by some superior theory, whose form is as yet unknown, but which itself is merely another approximation. However, if nonlinearity is false, then there will be every reason to believe that general relativity is really fundamental. Despite many words that have been written on this subject, there is nothing either in the field equations or in physical observation to suggest that general relativity has to be nonlinear. It is a crucial issue that cannot be determined simply on the basis of assumption.

In effect, in a nonlinear interpretation, we could imagine that if gravity curves space-time, then this curvature might act as a further source of gravity, and so create a situation in which, in a very strong gravitational field, *positive feedback* causes the gravitational effect to quickly get out of hand. This is what many imagine must happen in the interior of a black hole. The black hole idea in fact long predates relativity. It is a simple consequence of Newtonian gravity. For any body of mass m and radius r, there is a particular velocity which must be exceeded if anything is to escape from its surface. This velocity depends only on the ratio m/r; it can, in fact, be calculated at $\sqrt{2Gm/r}$. At the Earth's surface, we can calculate this as $\sqrt{2 \times 6.67 \times 10^{-11} \times 5.97 \times 10^{24}/(6.37 \times 10^6)} = 11,180$ metres per second or about 11.2 kilometres per second. Fortunately, we have the rocket technology to reach these speeds, so we can launch satellites and spaceships from Earth. If we were on the surface of Jupiter, however, where the mass is 1.9×10^{27} kilograms and the radius 6.68×10^7 metres, we would need to reach a speed of more than 58 kilometres per second to escape, a considerably more difficult proposition. Now we could imagine an astronomical body or perhaps large-scale system, such as a galaxy, where the escape velocity is greater than that of light (300,000 kilometres per second). Then even light couldn't escape.

This is purely classical gravitational theory, and follows a kind of reasoning dating from as early as 1783. However, according to many interpretations of general relativity, a body of this kind would also be sufficiently dense to initiate a positive feedback mechanism, leading to a complete gravitational collapse within the black hole, and create a state of *infinite density* at the centre. Now, remarkable objects have been discovered at the centres of galaxies which are very dark and of the order of a million solar masses. It may be that they fulfil the classical black hole condition, though no body of mass and radius sufficient to be described as a classical black hole, with $m/r = c^2/2G$, has as yet been discovered, with one remarkable exception — the universe itself! The universe that we see, with distant galaxies apparently moving away from us with speeds proportional to their distances, has an 'event horizon', a limiting distance (r) where the apparent recession velocity reaches the speed of light. It is now known that this observable or 'Hubble' universe has a mass density at the *critical value* where the event horizon condition, $c^2/2 = Gm/r$, is equivalent to the classical black hole condition, $m/r = c^2/2G$.

Interestingly, the Hubble universe has not yet collapsed to an infinitely dense singularity and even if we were to establish that objects like those at the centres of galaxies fulfil the same black hole condition, this would not be evidence that they were in a state of infinite gravitational collapse within. However, the nonlinear aspect of general relativity has been invoked to suppose that even bodies that don't reach the critical value of m/r could be dense enough for a positive feedback effect to enhance the gravity sufficiently to over-come all other forces, and so ensure that it must collapse below this value and *create* a black hole. Bodies of such high density *have* been discovered and *assumed* to be black hole singularities, based on the supposition that they would be if the nonlinear interpretation were correct. However, this is not independent evidence that the nonlinear interpretation *is* correct, and to assume that it is simply creating a self-fulfilling prophecy. The justification of the theory then works by the same kind of feedback loop as the theory itself.

In fact, the only way to use the effect as evidence would be to observe it actually happening. Observing separately either a large dense body or one with an m/r ratio greater than the critical value would not establish the process of black hole creation by nonlinear positive feedback because both could exist separately within the classical theory. Nonlinearity is very much the most problematic aspect of the theory, introducing singularities, violation of conservation laws, and many other anomalies. It is the most difficult aspect of general relativity to reconcile with other areas of physics, and it is one of the main reasons why a theory of quantum gravity has not yet been established. In view of this, we should always remember that the field equations of general relativity are an abstract mathematical structure with no direct connection to the physical assumptions from which they were derived or the interpretations which people have subsequently added, and beyond those effects which are definitely provable by experiment, they don't commit us to supporting these assumptions and interpretations.

Another aspect of general relativity that we should understand more fully before we rush to conclusions about its consequences is the idea that it proves that space-time is 'actually' curved, that its geometry is intrinsically non-Euclidean. We need to realise here that curvature is a mathematical technique for incorporating the effects of force and energy into a space-time coordinate system. If we have a field from any force, we can incorporate this into the equations of either classical or quantum physics by adding potential energy terms to the differentials in space and time. We call the modified system of differentials the covariant derivative. In principle, we have effectively changed the coordinate system; but while our x, y, z, t have acquired additional terms which modify their effects, our fundamental space and time structure has not changed. However, we could, instead, have said that the space-time combination was curved, and this would similarly require adding terms to the differentials. The effect would be the same, though this time the underlying structure would have changed. Exactly such a procedure has been done for electromagnetism and other forces, even for *Newtonian* gravity. In addition to this intrinsic ambiguity, there is also a danger of using curvature as 'explanation' rather than as effect, for it produces a level of finality without any obvious way of penetrating further. We are in danger of replacing physics with 'archaeology'. General relativity was constructed with the philosophy of removing the force of gravity from the picture and replacing it with a curvature in space-time. The successful use of the mathematical structure does not prove in itself that this philosophy is valid.

Cosmological equations for contracting, stationary and expanding universes were found by Alexander Friedmann in 1920 from general relativity. However, it is now known that exactly the same equations can be found quite easily from Newtonian theory, and that they are not dependent on the curvature of space but on the density of matter, whether this is interpreted as causing curvature or not. In fact, even if we interpret high or low density as indicating positive or negative curvature, the most recent evidence indicates quite clearly that the universe has exactly the *critical density* required to make it completely flat, that is completely Euclidean with no curvature at all. This is a result that certainly surprised the many who a few years earlier would have put their money on a closed, curved universe, emerging from something like a black hole-type singularity and no doubt eventually collapsing into something like the same kind of state. Space on a universal scale refused to be curved despite the almost universal belief at the time that it would be. The evidence also suggests that the universe may well be infinite with a great deal more of the same beyond the Hubble radius, which is the 'event horizon' or limit of observation beyond which the expansion rate exceeds the speed of light. A flat universe means that ours cannot be a 'walled garden' in the way that a curved one could, which has implications for certain interpretations of quantum mechanics as well as general relativity. We have learnt by successive enlightenments that the solar system isn't our universe, that the galaxy isn't our universe, and that even our 'universe' isn't our universe!

Chapter 3

Quantum Mechanics

3.1 Quantum mechanics

Quantum mechanics began to emerge in Göttingen 1925 when the young Werner Heisenberg was trying to make sense of the data on atomic spectra, that is, the radiation emitted when atoms are excited by a stimulus such as heat or an electrical discharge. The quantum theory, first devised by Max Planck in 1900, said that energy could not be transferred continuously between systems or within systems but only in discrete packets, which he called quanta. Pure energy is emitted from physical systems as electromagnetic radiation, which is wave-like with wavelength λ (distance between successive crests) and frequency ν (number of crests per second) multiplying to give the velocity of light ($c = \nu\lambda$), and includes gamma rays, X-rays, ultraviolet, visible light, infrared, microwaves and radiowaves. Planck showed that this radiation came in packets of size $h\nu$, where ν was the frequency and h a universal constant which was the same for all types of radiation. Einstein, in 1905, showed that the quanta had a particle-like aspect when he explained the photoelectric effect as a result of single high energy quanta — or, as we now call them, photons — striking single electrons within a metal with enough energy to allow them to escape from the metal surface.

In 1913 Niels Bohr addressed a problem that had arisen with the Rutherford model of the atom in which a cloud of negatively charged electrons could be imagined in a series of orbits around a tiny positively charged nucleus. Bohr said that the possible electron orbits for any given type of atom had energies fixed by their radii, and emissions and absorptions of radiation took fixed values of $h\nu$ as transitions took place between them. The emitted radiations would occur at certain definite frequencies, which, by scanning across the entire range of possible values, would be observed as a spectrum with a nonzero signal only at the allowed values. He found that this gave excellent agreement with the results obtained for the hydrogen atom, but the agreement was less good for heavier atoms, and theorists spent the next twelve years trying to adapt the theory in different ways to get a better fit to data.

Eventually, Heisenberg decided that Bohr's physical electron orbits with definite radii were incompatible with the data on frequencies obtained from spectra. He decided to abandon the hypothetical radii and retain the frequencies which could be observed and create a new mathematical structure. He would consider only the *observable* quantity, frequency, in constructing his theory and consider only the transitions between one 'energy state' of the atom and another in working out what frequency would be emitted or absorbed. Heisenberg's approach was very abstract, but within a year, a different, and seemingly more physical, way of looking at the same problem was arrived at by Erwin Schrödinger. Schrödinger wrote down an equation for an electron within an atom based on an extension of a theory of electron waves put forward by Louis de Broglie a few years earlier. His equation looked similar to a classical equation, except that physical quantities like energy and momentum became instead mathematical 'operators' (based on differentials with respect to space and time) acting on a wave-like object, called the 'wavefunction'. Starting with the classical equation, you could 'quantise' by replacing the energy and momentum terms (E, p_x, p_y, p_z) with differentials in the conjugate variables, time and space ($E \rightarrow i\partial/\partial t$; $p_x \rightarrow -i\partial/\partial x$; $p_y \rightarrow -i\partial/\partial y$; $p_z \rightarrow -i\partial/\partial z$) (where we have set another fundamental constant $h/2\pi$ equal to 1) and choosing a wavefunction (symbolised by ψ) that gave the energy and momentum values back when you applied the operators to it. In fact, you could even leave the quantum equation looking very much like the classical one by reinterpreting the symbols as operators rather than as quantities.

The wavefunction had some of the characteristics of a classical expression for a wave — an amplitude, related to the maximum 'height' of the wave, and a phase factor, a function which showed the variation in 'height' over space and time as the wave went through its cycle — but it had a different physical interpretation. In the quantum realm, the wavefunction appeared to be not a definite statement of the position of, say, an electron in a given time, as would be expected with a classical wave, but a mathematical expression related to the *probability* of finding that electron in that position at that time. The electron didn't actually have a position — there were no actual orbits round a nucleus, only a probability density (calculated by multiplying the wavefunction by its complex conjugate). By implication, this extended to the whole of matter, though on the large scale, quantum effects would be too small to be noticed. No material object had an actual position in space. It had a finite probability of being anywhere in the universe. The position where we imagined it to be was simply the most probable place of finding it, not its true position.

The original versions of quantum mechanics were not relativistic, though it was known that relativistic effects were significant at the quantum level. The search for a relativistic quantum mechanical equation for the electron proved to be difficult because it required the introduction of mathematics never before used in physics and produced results which at first seemed rather strange, but the Cambridge physicist Paul Dirac finally succeeded in discovering the required equation in 1927. Because it applies to all the truly fundamental particles in nature, the fermions, and because in nature at the fundamental level, there are only fermions and the interactions of fermions, this equation must be considered the most important in physics, and it is, in fact, the only equation featured on a stone in Westminster Abbey, where it appears in the form $i\gamma \cdot \partial\psi = m\psi$.

However, producing this most fundamental equation was very hard won, as it required even more radical changes in physics and its representation in mathematical form than the nonrelativistic quantum mechanics of Heisenberg and Schrödinger. The Dirac equation was based on operators and a wavefunction. However, the wavefunction was not a scalar wavefunction like Schrödinger's, but a *spinor* wavefunction with four simultaneous solutions, two of which required negative energy values, in contradiction to all previous observations.

It turned out that this was a prediction of antifermions, which were discovered a few years later. The equation also, for the first time, incorporated electron spin as a natural outcome. This was an intrinsic angular momentum of the electron necessary to explain the magnetic properties of most materials. The spin of the electron, however, was a very unusual form of angular momentum, for it required the electron to rotate round a complete cycle *twice*, or $2 \times 360° = 720°$, before getting back to its starting position. All fermions had this spin $\frac{1}{2}$ value whereas bosons, such as photons, had spin 1, with a 360° cycle. The spin was always quantised and came in units of $h/2\pi$, the Planck unit of angular momentum. It had just two directions, left- or right-handed with respect to the direction of motion of the particle, or more generally, spin 'up' or spin 'down'. The four terms in the Dirac spinor then represented the four combinations of particle and antiparticle, and two directions of spin. All had to be present even if the particle was defined as an electron, rather than an antielectron.

Another aspect of this theory was that an electron did not exist on its own. It interacted with some kind of unlocalised field, called 'vacuum', splitting it into 'virtual' (unobservable) electron–antielectron pairs, which produced further virtual pairs, and so on. To accommodate this concept, quantum mechanics had to be extended into a new range of mathematical structures, called quantum field theory, though the ultimate basis of these is still a form of the Dirac equation. In 1930, Schrödinger found a solution of the Dirac equation which gave the 'equation of motion' of a free electron, that is an equation which gives the particle's changes of position with time. He found that the equation had all the usual terms expected in classical physics (although in a quantised form), but with an additional term with no classical analogue. It turned out that this term predicted an additional violent, 'jittery motion'. We still use the German word *zitterbewegung*(meaning 'jittery motion') which Schrödinger introduced to describe it. This is because it is unlike any known motion in classical physics. It is now interpreted as a motion in and out of 'vacuum', a switching between the four states in the spinor. If the *zitterbewegung* didn't exist, an electron would travel at the speed of light, but the zigzagging in and out of vacuum slows it down to the speed observed. In relativity theory, a particle such as the photon which travels at the speed of light has no rest mass, but a particle with a rest mass travels slower. So, the *zitterbewegung* is in a sense the origin of the rest mass of the electron, and in fact, of any fermion.

Not only did the wavefunction of the Dirac equation require a new form, so also did the operator. The Schrödinger equation had a squared version of the space differentials and a linear (nonsquared) version of the differential in time. To make them both linear, Dirac had to introduce a complicated new algebra based on 4×4 matrices (2-dimensional arrays of 16 numbers). There were five of these so-called gamma matrices: four were included in the equation, and the fifth was needed to complete the algebra. When all possible multiplications of the matrices were carried out, there were a total of 64 terms in the algebra. The introduction of the matrices meant that though the equation was relativistic in including space and time as four dimensions at the same level, it was not quite 'purely' relativistic in the sense of Einstein's classical special theory. This has sometimes been seen as initiating the clash between quantum mechanics and relativity which has become increasingly apparent with the success of quantum mechanics. It was a union of two theories but required modifications to each. General relativity also required modifications to special relativity, but seemingly in the opposite direction. In principle, relativistic quantum mechanics required a loosening of the Minkowski concept of a 'union' between space and time, in that the two quantities were linked to different gamma matrices with time showing every sign of being unobservable, unlike space, while general relativity required a tightening of the union towards defining the two parts as indistinguishable components of a single physical quantity.

3.2 Is quantum mechanics a problem?

Quantum mechanics is the most successful physical theory ever devised. Despite its reputation for being based on probabilities rather than certainties, it has been responsible for the most exact predictions ever accomplished. The quantity known as the magnetic moment of the electron, for example, has been calculated and shown to agree with experimental results to an unprecedented 11 significant figures. In addition, quantum mechanics is the basis for the revolution in technology which has had such an impact on all our lives in the last half century; modern electronics, computing and laser technology, for example, are unthinkable without it. However, popular accounts and statements by many prominent physicists suggest that they are not entirely comfortable with it. It is described as 'weird' or 'strange', and regarded by many as a device for calculating results that turn out to be true when we do the measurements, but with no easily identifiable physical meaning. Most practitioners just shrug their shoulders and say this is what physics turns out to be like at the micro-level. It's not what we expected and it's not what we wanted, but it's what we have to live with. Maybe at some time in the future someone will find a better underlying theory which is more to our liking, but at the moment the signs are not hopeful.

Meanwhile a whole industry has developed on explaining this to the nonscientific public. Physicists — many of them very famous — have lined up in their thousands to proclaim that they don't understand it themselves and can't understand how anyone *could* understand it. So we have Niels Bohr saying "Those who are not shocked when they first come across quantum theory cannot possibly have understood it." Richard Feynman with "I think I can safely say that nobody understands quantum mechanics." John Wheeler giving his opinion that "If you are not completely confused by quantum mechanics, you do not understand it." And Roger Penrose categorically stating that: "Quantum mechanics makes absolutely no sense."[4] Of course, the mystery only adds to the magic and excitement of doing cutting-edge physics, making clear how far physics is from the routine kind of systematic investigation that many people wrongly assume science to be. The 'quantum' concept has even entered popular culture as an idea that suggests the breaking up of old patterns and old certainties. The "quantum leap" has become the metaphor for a massive change in understanding via an abrupt transition, even though in physics a quantum transition is the smallest possible change that can occur in nature!

However, quantum mechanics is far from new. It made its first appearance in 1925 in the work of Werner Heisenberg and has been in constant and quite rapid development ever since. It has produced a vast literature and is a required starting point for nearly all the major fields of physics research today, such as particle physics, nuclear physics, condensed matter physics and optics, not to mention quantum chemistry and a huge range of modern technologies. Numerous experiments have shown that the 'strange' effects it produces — superposition, nonlocality, entanglement, *etc.* — are valid and are not explained by any more traditional approach. So, why is there any problem with it today? We have long since ceased to find it a problem that an object carries on moving in a straight line even without a force, even though the idea is completely counter-intuitive and would not have been understood for many centuries after Aristotle. So, what is different about quantum mechanics?

One thing that certainly makes quantum mechanics difficult to be regarded as a physical idea rather than as a purely 'calculating engine' is the rather impenetrable nature of the mathematical apparatus it uses. In the first place, most treatments of quantum mechanics package all the information available about a system into an object called the wavefunction (usually symbolised by ψ), which acts as a kind of black box. We do operations on it and get experimentally testable results for particular conditions, but what it *is* is never properly specified — we have no way of knowing what is in the box. We know that the wavefunction has wavelike features such as 'amplitude' and 'phase', and that quantum mechanical equations are in some ways similar to the ordinary wave equations generated by classical physics, but the quantum mechanical wavefunction is most definitely not a description of an ordinary wave. In addition, the more advanced forms of quantum mechanics, such as relativistic quantum mechanics and quantum field theory, tend to introduce seemingly arbitrary mathematical complications, such as matrices, which destroy the elegance and symmetry that we have come to expect of supposedly fundamental physics.

The methods are old and well established and they yield good results, so we have developed a kind of 'recipe book' for quantum calculations which is exactly what we teach as quantum mechanics to our students. There are lots of things to calculate, so we haven't got the time to work out what is going on. A few people worry about the foundations of the subject, but that kind of thing can be left to the philosophers. Although some of the founders of quantum mechanics — including Bohr, Heisenberg and Schrödinger — were deeply interested in philosophy and even influenced by it, the kind of education they received is not common nowadays, and partly due to their efforts, there is far more physics for a student to absorb than there would have been in 1925. So, physics as calculation and physics as philosophy have gone their separate ways, and a market-driven world can't really afford the luxury of the latter.

However, the notion that trying to understand quantum mechanics is 'philosophy', rather than physics, is deeply mistaken. It is only by understanding the physics we have now that we can progress to the next stage. At the moment, this is impossible for two reasons. The first is that we have repeatedly refused to accept that quantum mechanics could represent any form of intrinsic truth. As long as we continue to think that it is 'weird' or not what it 'should' be, we will create a barrier between ourselves and any more fundamental theory. It is obvious from the success of quantum theory that we should be accepting nature at what is clearly its own valuation rather than our own. The second reason is that the mathematical structures and methods we have used to develop quantum theory, though ideal for calculation, seem almost deliberately designed to prevent us from understanding the physics at any deeper level. Contrary to what has been said by many commentators, neither of these problems is insurmountable. If we look at quantum mechanics from a more fundamental point of view, it quickly becomes apparent that it is not in any sense 'weird' or illogical, though this description could be applied to our naïve expectations. At the same time, the mathematical structures that we have so far used for quantum mechanical equations are by no means the only ones that we could have used; they were simply the ones that came about through the accidents of history, and there are others that give us a far better idea of how it relates to the rest of physics.

3.3 Is quantum mechanics weird?

Anyone who studies quantum mechanics at any level quickly comes across the fact that it produces many results that are entirely different to what we would expect from any previous experience of physics. One of the key aspects is that events and measurements are all defined in terms of probabilities. Exact knowledge is not just difficult to achieve because of our limitations as observers, but is actually intrinsically meaningless. Radioactivity is now a relatively well understood phenomenon. It is interesting that the three main types of radioactive emission — alpha, beta and gamma decays — represent phenomena which are as different as it is possible to be in the whole of nature, as they represent three wholly different physical forces. Radioactive decay generally follows an exponential law. This means that there is a given time in which approximately half of the atoms in any given sample will decay. It could be anything from a tiny fraction of a second to billions of years. After twice this 'half-life', half of the half remaining will decay, and after three times the half-life, half of the remaining quarter. The thing that remains unknown is exactly when any particular atom will decay. The process is purely statistical.

One particular kind of radioactive decay involves another typical quantum process. Alpha decay involves the formation of a grouping of two neutrons and two protons in the nucleus of a heavy atom, the resulting particle is known as the alpha particle; the grouping requires less energy than the four particles would need for acting separately, and so the alpha particle acquires this as kinetic energy. The energy, however, is not enough to allow the alpha particle to escape from the nucleus entirely. Nevertheless, some alpha particles do escape by a process which is called 'quantum tunnelling'. In effect, the energy possessed by the alpha particle and the energy needed to escape are only the *most probable* values. There is always a small possibility that an alpha particle could be on the other side of the 'energy barrier', which is what we mean by saying that it has 'tunnelled' its way through. It is the quantum mechanical way of saying, "And with one bound he was free." With billions of atoms in a sample, some will always be found to decay. No quantum process involving an energy exchange occurs in anything but probabilistic terms. There is always a finite chance, though it is often remotely small, of virtually anything happening. This sometimes causes misunderstandings when scientists interact with the media. Physicists will rarely say that X (say the creation of black holes in a large collider) couldn't happen; they will say that it has a very low probability. It may be such a low probability that it would take many universe lifetimes to get one occurrence, but it is still finite.

Another problem existed well before quantum mechanics was established, but the quantum explanation added extra complications. This is called wave–particle duality. All physical events involve a transfer of energy. There are two ways in which this can happen, either by continuous waves or through a stream of particles, like bullets. From early in the nineteenth century, we have had a sure-fire way of detecting the presence of waves. Thomas Young set up a double slit experiment, in which light was split into two beams through two narrow slits (originally pinholes) and then projected onto a screen behind. With modern light sources, especially those of a single colour, this is fairly easy to do. The result is a series of light and dark bands or fringes. Where the peaks and troughs coincide respectively, the intensity is high: the two waves enhance each other. This is called constructive interference. Where peaks of one wave coincide with troughs of another, the intensity will be low; this is called destructive interference. The differences in phase between the two waves changes with the angle of projection, and as the light spreads in a broad band on the screen, we see alternating regions of constructive and destructive intereference or bright and dark fringes. Young showed that the same effect was seen with water waves in a ripple tank, and we can also detect it with sound waves from loud-speakers (Figures 6 and 7).

However, there are other occasions on which light seems to behave like a stream of particles, now called photons. A classic one was the photoelectric effect investigated by Einstein in 1905, where a single photon seems to release a single electron and will only do so if it has a high enough frequency to produce the energy required. Eventually, it had to be accepted that light had a dual character, behaving

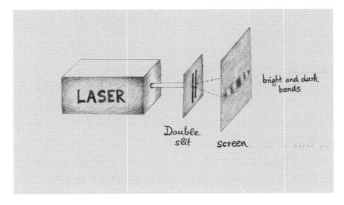

Figure 6. Interference pattern of light and dark bands produced by light from two slits.

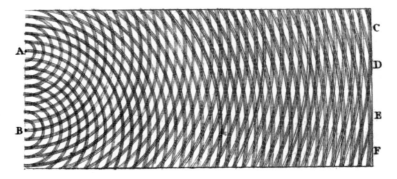

Figure 7. Young's original diagram showing how light bands are formed at C, D, E and F.

like waves in some circumstances and particles in others. Then, in 1924, Louis de Broglie speculated that objects always thought to be particles, like electrons, also had a wavelike character. He said that the particle-like momentum (p) generated a wave with a wavelength $\lambda = h/p$, where h is Planck's constant. It was quickly shown that electron waves did exist with the wavelike property of diffraction (spreading round an obstacle or through an aperture) and a new technology was developed which enabled waves with smaller wavelength than light to penetrate closer to the centre of matter using an electron microscope. The wavelike character of electrons and other particles (not to mention buckyballs and viruses) has now been categorically established, and it is now totally routine to do the Young's slits experiment with electrons.

Now, we can do a Young's slits experiment with electrons in such a way that we detect the electrons one by one as they emerge from the slits onto a target screen. If the slits are of the right dimensions, opening one slit alone will show a narrow strip of hits immediately behind the slit; opening the other slit will show a narrow strip of hits immediately behind that one; however, opening both slits at once will result in an interference pattern. Electrons are, of course, particles and so may be expected to go through one slit or the other. So, if they are sent through one at a time, they might be expected to create a narrow strip of hits behind each slit, but, in fact, they hit the screen in such a way that, over a period of time, the pattern builds up to form a set of 'bright' and 'dark' fringes, indicating regions of constructive and destructive interference.

The wave or particle behaviour seems to depend on what the experimenter decides to observe. If we have two slits open, electrons do not appear to choose one slit or the other but to go through *both slits at once*. The wave behaviour in quantum mechanics is linked to the idea that the electron doesn't have a position in space and time, only a probability of being in a given place at a given time. It isn't the electrons that go through the slits and interfere, but their probability distributions. Another way to put it is that the electron wavefunction is a superposition of two separate possibilities, going through slit A or going through slit B. It doesn't choose one or the other. It has a probability of doing both at once. Where we have a superposition of different possible states, the outcome remains indeterminate until a measurement is made. The wavefunction then 'collapses' to the single value found by the measurement. So, if we close A, it will certainly go through slit B. The observer will have made a 'measurement' and the outcome will be decided. If both slits are open the decision has not been made. It is possible to do an experiment similar to Young's slits in which the choice of whether to observe the outcome from one or both slits is made only *after* the electrons have passed through the slits, but the outcome is still the same. The *observer*, by looking at the results from both slits or one, effectively makes the decision on whether the electron wavefunction still contains two possible outcomes or has collapsed to one.

In quantum mechanics, it would seem that two or more possibilities can exist in a superposition of states with the outcome only decided when a measurement is taken. Schrödinger was so perturbed about this problem that he created a 'thought experiment' which imagined a cat in a superposition of states of being dead or alive simultaneously (Figure 8). In Schrödinger's own words (as translated by John D. Trimmer): "A cat is penned up in a steel chamber, along with the following device (which must be secured against direct interference by the cat): in a Geiger counter there is a tiny bit of radioactive substance, *so* small, that *perhaps* in the course of the hour one of the atoms decays, but also, with equal probability, perhaps none; if it happens, the counter tube discharges and through a relay releases a hammer which shatters a small flask of hydrocyanic

Figure 8. Schrödinger's cat.

acid. If one has left this entire system to itself for an hour, one would say that the cat still lives *if* meanwhile no atom has decayed. The psi-function of the entire system would express this by having in it the living and dead cat (pardon the expression) mixed or smeared out in equal parts."[5]

Obviously, no one would believe that a real cat could exist in a superposition of life and death states — a living organism contains too many interacting components for the states of life and death to be a simple quantum mechanical superposition. Also, there is a good case for describing a cat, or even an automatic measuring apparatus, as much an observer as a human one. In view of the difficulties uncovered by the pioneers of quantum mechanics, Niels Bohr and his colleagues set up a way of understanding the subject, which came to be called the Copenhagen interpretation, after the university city where most of the work was done. Bohr and colleagues said the world must be divided between a quantum system and an essentially classical measuring apparatus, and it was only when the system interacted with the measuring apparatus that a definite event could take place. Further quantum evolution would follow with a new superposition of states, but change to a new defined state would only happen after the next measurement. In fact, it is possible to prevent a system from evolving by continually observing it in the initial state provided by the first measurement. This is called the quantum Zeno effect by analogy with an argument by the Greek philosopher Zeno of Elea that an arrow in flight could not be moving because it could not be observed to move at any instant.

The 'operationalist' view of quantum mechanics provided by the Copenhagen interpretation has given physicists a method of using it as a working tool without having to worry too much about what it really 'means', but the concepts of 'measuring apparatus' and 'observer' are somewhat problematic as they involve ideas outside the realm of quantum mechanics. There is, however, another way of looking at it. When we talk about an 'interaction', say when we use Newton's third law of motion, we often discuss it in terms of an interaction between body or system A, and body or system B; but really, even in the simplest of cases, system A is not interacting with system B, but with the 'rest of the universe'. Most of this will, of course, be system B, but a small amount will include an interaction with everything else, the entire environment. In quantum mechanics, there is an expression for this environment; we call it 'vacuum'. It is the interaction with this environment that causes a quantum system to 'decohere' (or fail to maintain its quantum status) or a wavefunction to 'collapse', whether or not it contains anything that we would define as a classical measuring system.

Quantum 'entanglement' is a term introduced by Schrödinger to refer to another disturbing property of quantum systems; it was referred to as 'spooky action at a distance' by Einstein. Like many of the novel aspects of quantum mechanics, entanglement requires a version of *nonlocality*, a term indicating that some kind of 'interaction' has taken place at a speed greater than that of light, and probably instantaneously. Both superposition and the formation of combination states of more than one particle are nonlocal processes, and entanglement is a result of superposition. In a typical case, two photons are created simultaneously with exactly opposite spins (which can be observed through their directions of polarisation or planes in which their waves are made to vibrate). This conserves angular momentum because the total, taking into account direction, is zero, as it was before the pair production. Of course, each photon exists in a superposition of spin up and spin down states, until a measurement is made. If we now measure the spin state of one particle we will 'collapse' the wavefunction and produce either spin up or down, let's say spin up. Simultaneously, there will be an *instantaneous* fixing of the spin state of the other particle to spin down, however far apart the photons are, and even when they are beyond communicating at the speed of light in the time allowed for the experiment. A considerable number of experiments, notably Alain Aspect's of 1982, have shown that such an effect is observable. It is the basis of the technology of quantum cryptography, now close to fruition. Quantum teleportation, the transfer of quantum information about entangled states over long distances, has now been achieved over 140 kilometres.

One of the earliest developments from the Heisenberg picture, the Heisenberg uncertainty principle, codified one of the reasons why Heisenberg had to begin by abandoning explaining the atom using unobservable quantities like orbital radius, but created a new problem of its own. According to this principle, the conjugate variables, energy and time, momentum and space (or position), cannot be known exactly at the same time. So, if the momentum of a particle p is known to an accuracy Δp, then the position of the particle x cannot be known to better than Δx, where the product of Δp and Δx is greater than or approximately equal to the constant $h/4\pi$. If one of these quantities is known *exactly*, then the other cannot be known at all. This strikes at the heart of classical physics, where the momentum and position of all particles combine to give the complete information on the evolution of a system. The uncertainty principle says that accurate physical knowledge is not limited merely by the inadequacy of the observer, but by the intrinsic impossibility of such knowledge.

One of the key ways in which the uncertainty principle acts is in the creation and annihilation of *virtual particles*. These are particles which never actually materialise because they are produced below the energy required for this to happen. The bosons used in interparticle interactions — strong, weak or electric — are usually virtual, though conditions can be set up in which they can materialise and so become 'real'. They are allowed to remain in a virtual condition because they have very short lifetimes. So they effectively 'borrow' the required energy by reduction in the amount of time.

Several other aspects of quantum mechanics have also been regarded as strange, at least by the standards of naïve realism. The spin of particles has several properties which distinguish it from conventional angular momentum, and it certainly doesn't appear to be due to a simple kind of rotation in space. We have already discussed the spin $\frac{1}{2}$ state of fermions and the associated *zitterbewegung*, and it is far from obvious (except from mathematical requirement) why a fermion requires four states, including two antistates, in its wavefunction. Another 'strange' property of fermions is the Pauli exclusion principle. According to this principle, no two fermions, anywhere in the universe, can have the same set of 'quantum numbers', *i.e.* the same energy, momentum, angular momentum, spin, *etc.* This principle is ultimately responsible for the whole of chemical structure, as it determines the energy states of electrons in atoms and molecules; it is also responsible for those electrical properties of condensed matter states (solids and liquids) which have led to modern 'solid state' electronics and computers. Very significantly, also, it seems to imply some kind of holistic connection between all the components of the universe, as well as an instantaneous (nonlocal) correlation between all fermions. How else can a fermion 'know' that its quantum numbers are different from those of all other fermions?

Numerous attempts have been made to find an explanation of quantum mechanics that is closer to traditional ways of understanding things. David Bohm took the Schrödinger equation, split it into two parts (real and imaginary), and found a term that acted rather like a classical potential energy term, but clearly had a quantum origin. This 'quantum potential' could be interpreted as 'guiding' the particle in a totally deterministic way, and so remove the probabilistic interpretation of a particle's position. The particles would have known trajectories given knowledge of the initial conditions; the uncertainty would come from lack of knowledge of these. However, Bohm's theory is still nonlocal, as the quantum field or potential pervades the whole universe and instantaneously connects all aspects of the system. In addition, it is nonrelativistic, derived from the Schrödinger and not the Dirac theory, and has not so far been extended in this direction. It gives an interesting 'physical interpretation' of aspects of quantum theory, but cannot be the final explanation.

Another way of looking at quantum mechanics, though this is more of a quantum formalism than an attempted explanation, is Feynman's 'path integral' or 'sum over histories' approach. Here, the particle is assumed to travel all possible paths simultaneously, however 'way out' any individual path may be. When the summation of all the paths is done, most of the possible routes cancel, leaving just the final route taken. This is similar to classical approaches based on the principle of least action, and is a powerful method used extensively in particle physics, but it doesn't really overcome the problems posed by probabilistic interpretations, entanglement and nonlocality.

An interpretation which has achieved a certain amount of support since it was first put forward in 1957 is the 'many worlds' interpretation of Hugh Everett. According to this, when any choice presents itself to a quantum system, as, for example, the spins of entangled photons, the universe actually splits into multiple universes representing the possible choices. So, for the entangled photons A and B, there are two overlapping universes, one in which A has spin up and B spin down, and another in which A has spin down and B spin up. However, when the measurement is made by an observer that, say, A has spin up, then the alternative universe splits off forever from the observer but carries on with its own evolution. Some would say that this is just a 'philosophical' interpretation of the mathematics of quantum mechanics, and that the fact that it includes no new observable features means that it contains no additional 'scientific' content. I think that it has two possible meanings. If you take it literally, then there really must be other worlds 'out there', perhaps with other versions of ourselves leading totally parallel lives. On the other hand, if you think the alternative universes are only 'virtual' worlds, then the whole concept is just a metaphor and has no additional meaning.

Perhaps one of the reasons for the popularity of the 'many worlds' view is the seemingly similar cosmological theory of multiverses, in which our own universe is one of possibly an infinite number which could well have different laws of physics and different values for physical constants. In an echo of the 'anthropic principle' — that the universe is constituted as it is because human observers are there to observe it — the multiverse theory proposes that our universe is one which allows observers to evolve to observe it, whereas one in which, say, the proton was heavier than the neutron could never create a chemistry or biology which would allow this to happen. The multiverse theory seems to be going as far down the road of abjuring any responsibility for doing physics as could possibly happen. If you accept its conclusions, you are saying that there is no 'physics', no subject which logically can lead us to understanding the universe on fundamental principles. It is a negation of all that has led physics to one success after another in its refusal to compromise on the idea that there are abstract principles that we have to believe are true in any place in any era. It is an impossible theory for a physicist to believe, but the fact that it has ever been given any credence can be taken as a symptom of the impasse which has hindered any real development in the last few decades.

3.4 Is classical physics weird?

The idea that quantum mechanics is 'weird' is partly based on the assumption that classical physics is not. In classical physics, we have 'real' objects — atoms, particles, stars, galaxies, *etc.* — that interact with each other with forces that take time to take effect. Originally, in seventeenth-century Newtonian physics, one force, gravity, was assumed to be instantaneous no matter how great the distance between the objects (though, since the mechanism was unknown, this was more a default option than a doctrine). However, since the arrival of Einstein's theory of relativity (and even before in the case of electromagnetic forces), it has been assumed that the transmission of the 'information' which makes the interaction possible cannot be faster than the speed of light. There are four known forces in nature — gravity, electromagnetism, and the weak and strong nuclear forces — and all but gravity are known to be limited by the speed of light in terms of information. In the case of gravity, the limiting speed remains an assumption and has not yet been established by a conclusive experiment.

An interaction that cannot be transmitted instantaneously is called a *local* interaction. Local interactions are an intrinsic component of quantum as well as classical physics. In the case of quantum field theory, which extends quantum mechanics to the direct consideration of interactions, the forces between fundamental particles of matter are carried by other particles which are called bosons or exchange particles. Only in the case of gravity does the boson remain undiscovered. In the local interactions, the bosons travel at a finite speed (the speed of light or less) and carry 'information' (or energy) from one particle to another.

In classical physics we have real components of tangible 'matter' occupying real positions in space at any given time. Each of these parcels of matter or 'particles' has a given momentum (measured as mass × velocity), which may change as the particle is subjected to interactions from the other particles, and if we know how the positions and momenta of the particles change over time, we have an entire description of the system. The key point is that the positions of the particles and their momenta are definite, whether we know them or not, and there is no instantaneous connection between particles in different localities, only one that is limited by changes transmitted at the speed of light or less.

Now, this seems to many people to be the 'normal' state of affairs. It is what we are familiar with from everyday experience and it explains many things that we can see with our own eyes and put to the test in simple experiments, such as the behaviour of balls on billiard tables. But from the point of view of fundamental physics, it is not normal at all. It is, in fact, seriously weird.

The most important fact about physics, the one that makes it the unique route to understanding the fundamental principles of nature, is that it is a one-way ticket. There is no going back. Once we set out on this route, it makes no sense to call a halt until we have arrived at the ultimate physical explanation, and if we have a picture of reality as made up of objects of 'real' matter operating in a semi-abstract space according to the flow of an even more abstract time, we have clearly not reached a final explanation. We are still operating on what we might call a 'storybook' level. We have at least three different types of concept with no fundamental connection between them. It is a bit like mythological stories in which gods and supernatural beings interact with humans, or those children's stories in which we find animals speaking to humans in the same language and operating as though they were humans.

There is nothing wrong with doing this. It is obviously served us well in the past. What is wrong is expecting that things will continue in this way as we get to deeper levels and more fundamental theories, and not taking the opportunity to go beyond it when nature seems to demand that we should. A theory in which we have separate concepts of space, time and matter, only partially abstract, is most definitely not a plausible fundamental theory. In fact, expecting it to continue is a kind of denial of what we have so far achieved. The history of physics at the fundamental level has been like the successive replacement of ideas that stem from concrete or empirical experience with generalising abstractions. The future of physics appears to be abstract. Only a totally abstract theory could make sense at a fundamental level. The presence of anything else is simply an indication of inadequate explanation.

We could, in fact, consider Einstein's general theory of relativity as an instance of an attempt to unify at the 'real' or storybook level, at least in the way it is normally interpreted, and so it is perhaps not surprising Einstein also became the greatest twentieth-century critic of quantum mechanics, despite his own significant contribution to the development of the quantum concept. General relativity not only incorporated time as a fourth dimension added to the spatial structure, but also eliminated gravity and the effect of mass by stating that the whole structure could be curved. So, as we have seen, if the space-time in the vicinity of the Sun was sufficiently curved, an object approaching it would follow the lines of curvature and appear to be attracted to the centre as in gravity.

Creating space-time and making it curved to include the effect of matter, though possible mathematically and useful to some extent as a calculating device, doesn't give us any way of understanding the differences between space and time, which we definitely observe in quantum mechanics, and between these and matter. It creates a position beyond which we are unable to go, and yet doesn't explain many of the parts or the relationships. It reminds me of doing a puzzle in which you fit the pieces together and it looks partially persuasive but you know something is wrong. Despite making a lot of headway, you eventually reach an impasse and have to start all over again. You can't modify what you have done to get over the problem. As we know, we can't fit the pieces together in a classical unified field theory, as Einstein had once hoped. Quantum gravity has not delivered successful results either. We need space, time and matter to become part of a unity which isn't a forced one.

There is an even greater problem that even if we could succeed in integrating all the forces into a unified picture using a space-time combination with multiple curvatures, we would be left with something at the end that itself had a complicated structure — exactly the thing that we had previously always avoided when forming a new theory. The key to success out of the maze is always to leave an opening for the next level. If we create a closed but complicated structure, there is no opening.

Chapter 4

Simplicity and Abstraction

4.1 Keep it simple

The lesson that we seem to learn from studying the history of physics is that nothing works so well as keeping it simple. Unfortunately, this is easier said than done, for it turns out that our 'simple' and nature's 'simple' are quite different concepts. We might think of a table (to use the philosophers' famous example) as a 'simple' thing, but to nature it is anything but simple. Even if we were to convert every atom in the observable universe into paper and ink, we still couldn't write down an exhaustive description of the object. Eddington famously described this in his *The Nature of the Physical World* (1928), when he wrote of having 'two tables' before him. "One of them," he says, "has been familiar to me from earliest years. It is a commonplace object of that environment which I call the world." This table is "comparatively permanent"; it is also "coloured" and "substantial", that is, it doesn't collapse when anyone leans on it. Eddington says that, "if you are a plain commonsense man, not too much worried with scientific scruples, you will be confident that you understand the nature of an ordinary table. I have even heard of plain men who had the idea that they could better understand the mystery of their own nature if scientists would discover a way of explaining it in terms of the easily comprehensible nature of a table."

The other table is very different: "Table No. 2 is my scientific table. It is a more recent acquaintance and I do not feel so familiar with it. It does not belong to the world previously mentioned, that world which spontaneously appears around me when I open my eyes, though how much of it is objective and how much subjective I do not here consider. It is part of a world which in more devious ways has forced itself on my attention. My scientific table is mostly emptiness. Sparsely scattered in that emptiness are numerous electric charges rushing about with great speed; but their combined bulk amounts to less than a billionth of the bulk of the table itself. Notwithstanding its strange construction it turns out to be an entirely efficient table. It supports my writing paper as satisfactorily as table No. 1; for when I lay the paper on it the little electric particles with their headlong speed keep on hitting the underside, so that the paper is maintained in shuttlecock fashion at a nearly steady level. If I lean upon this table I shall not go through; or, to be strictly accurate, the chance of my scientific elbow going through my scientific table is so excessively small that it can be neglected in practical life. Reviewing their properties one by one, there seems to be nothing to choose between the two tables for ordinary purposes; but when abnormal circumstances befall, then my scientific table shows to advantage. If the house catches fire my scientific table will dissolve quite naturally into scientific smoke, whereas my familiar table undergoes a metamorphosis of its substantial nature which I can only regard as miraculous."[6]

Our 'simple' is really another way of saying 'familiar', so familiar that we can give a simple label to a complicated object. Nature's version of 'simple' tends to be an abstract category that we find extremely hard to grasp. So, while it is a well-known fact that the best ideas are often simple in this sense, they are equally often difficult to grasp, and generate resistance in the scientific community which can seem extremely strange in retrospect. The twentieth century had a long record of resistance to the most significant innovations in physics whenever they were simple or simplifying, very much to the detriment of the subject, and nothing much has changed since. Abstraction does not come easily to the human mind, which is more readily persuaded by the 'reality' of the concrete.

People often come up with 'unifying' theories that are dependent on a structural model related to the concrete world, the world of everyday experience, but nature seems to prefer unifications based on totally abstract ideas. In the mid-nineteenth century, as we have seen, James Clerk Maxwell came up with a unifying theory of electricity, magnetism and optics, based on just four abstract equations, which were also highly symmetrical. In fact, he added an extra term to make them so, and so discovered electromagnetic waves. However, Lord Kelvin, the most celebrated physicist of his day, could never accept it, and his *Baltimore Lectures* explain why: "I never satisfy myself until I can make a mechanical model of a thing. If I can make a mechanical model I can understand it. As long as I cannot make a mechanical model all the way through I cannot understand; and that is why I cannot understand the electromagnetic theory."[7] What Kelvin wanted was some kind of model in which a kind of fluid, called the *aether*, filled space and, working on good mechanical principles, gave objects their electrical and magnetic properties. Though Maxwell's theory eventually won out and all the many theories based on mechanical aethers with special 'physically understandable' properties were gradually abandoned, Kelvin's is still very much the attitude, conscious or unconscious, of many physicists given the choice between model-building and pure abstraction. It is extremely difficult to 'think simple' in the way that nature seems to require.

4.2 Abstraction

An unprejudiced view of physics would say that the most powerful and general theories have always shown a strong tendency towards abstraction. It is this abstraction that makes them apply to more than one particular set of circumstances — in fact, this is almost a definition of abstraction. The most powerful ideas we have ever had for understanding the world are the concepts of space and time, which seem to be almost hard wired into the human brain. Philosophers have debated for centuries, even millennia, whether these are illusion or reality, but certainly no method of constructing even a 'storybook' picture of the universe seems imaginable without them. In the Middle Ages, there was a major advance when theologians saw the abstractness of space and time as the nearest way to approaching (in Stephen Hawking's phrase) the "mind of God". A particularly powerful group of late mediaeval thinkers began to treat these concepts in exactly

the way that would lay the groundwork for modern physics. Their
mathematisation of space, time and motion, with a hint of the calculus that followed later, was a great practical step forward. They were
particularly concerned with *kinematics*, or the study of the movement of matter in space over periods of time, without regard to how
the motion originated. They considered both uniform and 'difform'
motion, or motion under constant velocity and motion with acceleration, including the very important case of motion under constant
acceleration. The culmination of this kinematical tradition came with
the work of Galileo who in the early seventeenth century applied it
to the case of bodies falling from heights close to the Earth's surface.
The problems brought up by the kinematic approach and the change
of one variable quantity with respect to another required the introduction of a new kind of mathematics, the calculus, which emerged
in the seventeenth century partly under the direct stimulus of solving
the physical problems.

A very significant subsequent development was the introduction
of mass as an abstract quantity at the same level as space and time in
the work of Newton. Mass was such an abstract notion that Newton
found it difficult to give it a 'physical' definition, other than the
rather circular one of the 'quantity of matter' in a substance. The
introduction of mass meant that kinematics now became dynamics, in
which causes of motion were as important as the motions themselves.
It also meant that physics had to become universal in scale, as mass,
unlike space and time, was a conserved quantity, and its introduction
into equations at the same level as space and time led to the creation
of new composite quantities, built up through combinations of various measures of mass, space and time, which followed universal conservation laws — for example, momentum, angular momentum and
energy. In addition, mass, through gravity, became a source for the
changes observed in nature, rather than just a means of measuring
their effects.

Mass, however, though seemingly part of the 'tangible' rather
than purely abstract aspect of nature and seemingly associated with
identifiable objects, was abstract in being a property of points in
space in the mathematical equations. Newtonian laws were defined

for point-like masses or for masses acting at a single point rather than for extended objects. To a large extent, the mass concept could be divorced from the object with which it seemed to be associated. Also, masses, in the case of gravity, appeared in the equations in squared form, as did space and time. Though the concept of mass has since been revised to incorporate the dynamic behaviour of objects as well as their so-called rest mass, this, if anything, has made it seem even less 'tangible' and more abstract. As we have gained a deeper understanding of the forces of nature and the quantum picture, even the 'rest mass' has come to seem increasingly dynamic in origin. The particles of 'tangible' matter do seem to be just points in otherwise empty space rather than objects with solidity and extension, and the masses associated with them to be a result of dynamic behaviour of one kind or another.

After gravity had given us the clue to the large-scale motions in the universe, other forces were found to be responsible for events on smaller scales. Eventually the electrostatic force known since antiquity was shown to be of the same kind as gravity, with an inverse square law named after Coulomb who first established it by direct measurement, but powered by a new quantity called charge, which was equally as abstract as mass. If anything, it was more abstract, for charge had no other physical function apart from its existence as a source of the electric force, while mass had an additional dynamic existence as a measure of the resistance to motion. Ultimately, all the effects of electricity, magnetism, optics, chemistry, biology, surface tension, capillarity, cohesion and friction could be attributed to the presence of this quantity. Since then, two other forces have been discovered, and though they are very different in many ways from the electric force, they do seem to originate from a very similar charge-like parameter. Though it does not seem to have been usual in the early days of the Standard Model to refer to these quantities as 'charges', this usage has now become gradually accepted, and with it, the possibility that, at least under the energy conditions under which the forces are thought to become unified (Grand Unification), they are aspects of a single parameter. No other quantity in physics seems to have reached the fundamental level provided by space, time, mass and charge, which have the status of being the only known means of apprehending events in nature and the proposed sources for these events. Other quantities used in physics equations seem to be in some sense composites constructed from these, and a method called dimensional analysis has long been used to work out these constructions.

Quantum mechanics suggests that we have been too timid in applying abstraction in physics. 'Tangible' reality is just an illusion. It is clearly wrong to try to construct a fundamental physics in which abstract ideas relate to tangible objects — a world like this would mean we could never get to the beginning. Obviously, we had to do this in the early stages of the subject, and it was a long struggle to create generalising abstract principles from empirical observations. But we need now to take the process to the final stage. Quantum mechanics tells us that we should not be separating out the supposedly 'tangible' and 'real' from an abstract construction applied to it, but treating all aspects of physics at the same abstract level.

4.3 Missing an open goal

Undoubtedly, the most fundamental questions are concerned with space and time. They always have been, as these are the only ways of directly apprehending the world around us. One of the most fundamental questions must be: why is this world 3-dimensional in space? This has always been considered a major problem, perhaps insuperable. It is certainly a problem if you, say, start with 10 or 11 dimensions as in string/membrane theory, and try to find out how they can be reduced to the ones we actually observe. But perhaps it shouldn't be, for we may well have long had the answer, or at least a very suggestive clue. The problem is not so much physics or mathematics as our history. We have for nearly two centuries had a very good indication of the reason. Unfortunately, this has never become general knowledge because the relatively simple mathematics required very early on acquired a bad name. Round about 1900, physicists missed an open goal and physics has never totally recovered from the disaster.

The key development was made as long ago as 1843 by Sir William Rowan Hamilton, who, though a mathematician, was employed as Astronomer Royal for Ireland. Hamilton started with complex numbers and the Argand diagram. We have seen how this can be used as a 'natural' representation of 2-dimensionality, even if this is slightly awkward. Can it be extended to 3? Can we draw another axis at right angles to the two on this diagram and, if we could, would it make mathematical sense? Clearly, it can't represent real numbers. They are all on the horizontal line and can't be anywhere else. If the diagram is to work like dimensions do in real space, then it would have to involve the addition of squared numbers via Pythagoras' theorem. Could it be another set of imaginary numbers? Here, we have a loophole. Though we can write down $i = \sqrt{(-1)}$, we have no idea what i actually *is*, or if there is only one kind of it. So we guess that there is another solution with, say, $j = \sqrt{(-1)}$, but with j totally distinct from i (Figure 9).

Now, any genuine algebraic system has to have the property of *closure* under multiplication. That is, if we do all possible multiplications of terms within the system, then nothing new should be produced. Complex algebra has this property. If I multiply $3 + 2i$ by $5 - 6i$, that is, multiply each term in the first expression by each term in the second, I will get the answer $15 + 10i - 18i - 12i^2 = 27 - 8i$. This is clearly another complex number, so it can be represented on the Argand diagram and I haven't produced anything new. Suppose, however, that I have a system made up of real numbers plus multiples of i and multiples of j. Then

real numbers $\times$ real numbers $\rightarrow$ real numbers;
real numbers $\times$ multiples of $i \rightarrow$ multiples of i;

Figure 9. Hamilton's attempt at creating triads.

real numbers × multiples of j → multiples of j;
multiples of i × multiples of i → real numbers;
multiples of j × multiples of j → real numbers.

All these are within the system. But when we try to multiply i's by j's, we have a problem. The answer can't be real because this would mean i's were the same as j's; it can't be a multiple of i because that would make j's real; and it can't be a multiple of j because that would make i's real.

Hamilton struggled with this problem for thirteen years. Eventually, he decided that the product of i and j had to be something new, say k, a *third* system of imaginary numbers, and 3-dimensionality had to be represented not by a combination of real and imaginary axes but by a system of three imaginary axes all at right angles to each other. The axes also had to be cyclic, rotating into each other by multiplication: multiplying i by j gave k; k by i gave j; and j by k gave i again. So, we can write the multiplication rules:

$$i^2 = j^2 = k^2 = -1, \quad ij = k, \quad jk = i, \quad ki = j,$$

which can be combined into the single expression:

$$i^2 = j^2 = k^2 = ijk = -1.$$

Here we have used a new symbolism (red bold italic) to emphasize that these new algebraic units have additional properties to the ordinary $i = \sqrt{-1}$ (Figure 10).

The discovery was a dramatic one, a classic example of the 'eureka' moment or flash of inspiration, made on 16 November 1843 while Hamilton was on his way from Dunsink Observatory, where he worked, to a meeting of the Royal Irish Academy in Dublin, where he was President. The account he wrote later for his son also hints at the

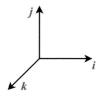

Figure 10. Hamilton's quaternions.

fact that Hamilton's marriage was not the most successful aspect of his life: "I was walking in to attend and preside, and your mother was walking with me along the Royal Canal...; and although she talked with me now and then, yet an *under-current* of thought was going on in my mind, which gave at last a *result*, whereof it is not too much to say that I felt *at once* the importance. An *electric* circuit seemed to *close*; and a spark flashed forth, the herald (as I *foresaw, immediately*) of many long years to come of definitely directed thought and work, by myself if spared, and at all events on the part of others, if I should even be allowed to live long enough distinctly to communicate the discovery. Nor could I resist the impulse — unphilosophical as it may have been — to cut with a knife on a stone of Brougham Bridge, as we passed it, the fundamental formula...."[8] The bridge still survives but there is no trace of the equations.

Hamilton knew it was not an ordinary discovery. He suspected what Georg Frobenius would prove, in 1878, that the system was unique, and that it was the key insight into the nature of 3-dimensional space. He had a track record for making large-scale powerful generalisations, which, though little appreciated at the time, would prove their worth when later generations caught up with them. His great system of dynamics, based on a remarkable analogy with optics, proved to be the only one which could be readily transformed into the new quantum mechanics a hundred years after it was first devised.

He thought up a striking name for his system based on the fact that the three imaginary parts needed completion with a fourth part (real numbers) for algebraic closure. With his well-known interest in poetry, he may also have been inspired by the lines from Milton's *Paradise Lost*, referring to the four ancient elements, air, earth, fire and water:

> Air, and ye Elements, the eldest birth
> Of Nature's womb, that in quaternion run
> Perpetual circle, multiform, and mix
> And nourish all things, let your ceaseless change
> Vary to our great Maker still new praise.
>
> (V, 180–184)

The fourth part of the *quaternion* could not be represented on the same diagram as the three imaginary axes, now standing for 3-dimensional space, but Hamilton almost immediately thought up a meaning for it, writing to fellow mathematician Augustus de Morgan on 9 December 1844: "My *real* is the representation of a *fourth* dimension, inclined equally to all lines in space." And to Reverend James Hamilton on 11 January 1845: "My letter related to a certain synthesis of the notions of time and space, or in the greatest abstraction of Uno-dimensional and Tri-dimensional Progression, the result being a Quaterno-dimensional Progression, or what I call a Quaternion." He even went so far as to express the idea in verse in a poem called *The Tectactrys*: "And how the One of Time, / of Space the Three, / Might in the Chain of Symbol, girdled be."[9] We can hardly fail to notice the similarity to the space-time concept used in relativity, and the idea that time might be a fourth dimension to the three of space is even older than this, but Hamilton now had a mathematical structure which *demanded* the connection.

The idea is so strikingly simple that you may wonder why Hamilton took so long to make up his mind that it worked. But the reason becomes apparent when we try multiplying out a term like $ij\,ji$. If we multiply out the two j's first, then, because $j^2 = -1$, we can reduce the expression to $-ii$, and because $i^2 = -1$, this becomes 1. So

$$ij\,ji = 1.$$

Now, as $ij = k$, this means that ji must equal $-k$. So

$$ij = -ji = k,$$

and we have, for the first time, established an algebra in which the order of multiplication matters. If we multiply the numbers 2 and 3, it doesn't matter whether we take 2×3 or 3×2, the answer is still 6. The same is true even for complex numbers. Their multiplication is *commutative*. However, for quaternions, reversing the order of the symbols also reverses the sign of the product. Their multiplication is *anticommutative*. Hamilton hesitated for so long because he had to break one of the then accepted laws of algebra: the law of commutative multiplication.

However, once we have introduced anticommutativity, it turns out not to be a problem at all, because space as we know it is actually anticommutative. If we turn a screw anticlockwise, we might expect it to screw upward (unscrew); this is like turning from x to y or i to j, and going in the positive z or k direction. However, if we turn the screw clockwise, it will go downward because we are rotating from y to x, and going in the negative z or k direction (Figure 11). Screw threads are, of course, produced in right- and left-handed forms, and the fact that this is possible is ultimately due to the anticommutativity of space. There are also right- and left-handed versions of many chemical molecules including a considerable number that are biologically important.

Again, an electric motor and a dynamo are constructed in the same basic manner to do reverse operations. If we place a rectangular coil between the pole pieces of a permanent magnet and send a current round the coil, one side of the coil will be thrust upwards and the other downwards as the current travels in opposite directions on two sides of the coil. The coil will then rotate. This is the principle of the electric motor, which produces mechanical motion from an electric current (Figure 12). Now, exactly the same apparatus can be used to make a dynamo or generator. This time, you rotate the coil between the pole pieces and generate a current. However, if we

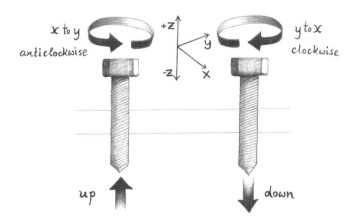

Figure 11. Anticommutativity of space demonstrated by screw movement.

coil turns

coil
rotated
mechanically

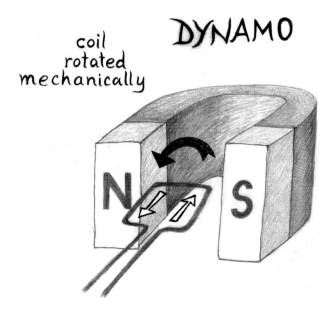

Figure 12. Motor and dynamo have opposite directions of currents in coil.

rotate the coil in a dynamo in the same direction as in a motor, with identical magnetic pole pieces, the current will be produced in the *opposite* direction.

A well-known mnemonic for the motor (Fleming's left-hand rule) has the thumb and first two fingers of the left hand arranged at right angles to each other. If the first finger gives us the direction of the magnetic field (north to south) and the second finger the direction of current, the thumb gives the direction of force or thrust. Exactly the same mnemonic can be used in the dynamo to work out the direction of current from the directions of magnetic field and thrust, but this time the *right* hand is used (Fleming's right-hand rule). Reversing the operation has reversed the sign because the operation is anticommutative, involving the product of two vectors at right angles (Figure 13).

The quaternions, as Hamilton suspected, were special. Anticommutativity only allowed three dimensions. It was the end of the line. Ignoring purely numerical values, if a and b are anticommutative, *i.e.* ab = −ba, then a, b and ab are anticommutative with each other but with nothing else. It is as though anticommutative things 'know'

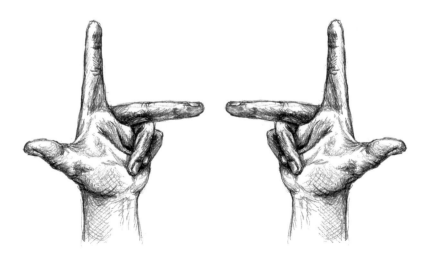

Figure 13. Fleming's left- and right-hand rules for motor and dynamo show different relative directions for current (second finger) relative to magnetic field (first finger) and thrust or mechanical force (thumb).

of each other's existence and form a closed off existence of their own, but commutative things are completely oblivious to the existence of anything else, whether commutative or anticommutative. Significantly, also, as we often want to subtract anticommutative products, if **a** and **b** are anticommutative, then, instead of **ab** − **ba** = 0, we have **ab** − **ba** = 2**ab**. Often, this is signified by the presence of an unexpected factor 2, or $\frac{1}{2}$ on the other side of the equation, and it has a particular relevance in quantum mechanics.

There is just one exception to the rule of three, and it was discovered by Hamilton's contemporaries. You can form a set of seven anticommutating square roots of –1, which, together with the real number unit (1), are called *octonions*, but these break another rule of algebra. They are *antiassociative*. That is, the *groupings* of terms multiplied matters as well as the order. So if you multiply a, b and c in that order, it still matters whether you multiply ab by c or a by bc. Antiassociativity means that you get results like $(ab)c = -a(bc)$. But with this single exception, there are no higher dimensional systems available to model systems like space. As was proved by Frobenius in 1878, the only division algebras that can exist are real numbers, complex numbers, quaternions and octonions.

Though he didn't prove the uniqueness of quaternions, Hamilton knew that he had discovered something of extraordinary value. He spent the rest of his life trying to develop his new system and its applications. A contemporary mathematician, Peter Guthrie Tait of Edinburgh, took up the cause and persuaded the great James Clerk Maxwell that he should use quaternions as an alternative formulation in his famous *Treatise on Electricity and Magnetism* of 1873. Here, it is interesting that Hamilton should use the analogy of an electrical circuit to explain his initial moment of discovery, for the vector representations of electricity and magnetism were precisely the kind of thing that quaternions could explain with ease. Maxwell's four equations of electromagnetism (as we now know them) become just one when we use quaternions. Maxwell didn't manage to do this personally, partly because he never reduced his equations to just four in the first place, and partly because the space-time connection which was crucial to the inclusion of magnetism somehow failed to be made.

But, in any case, Maxwell died young, and he had no time to make further contributions.

Nonetheless, everything seemed to be in favour of quaternions revolutionising physics; but it all went horribly wrong. Quaternions, as Hamilton believed, explained 3-dimensionality and suggested a link between space and time; they had to be the key to the universe. E. T. Bell, who wrote a well-known popular work on *Men of Mathematics*, however, proclaimed nearly a century later: "Never has a great mathematician been more hopelessly wrong." And he subtitled his chapter on Hamilton's career "An Irish Tragedy". Hamilton certainly made an unhappy marriage and later turned to drink, but, according to Bell, "Hamilton's deepest tragedy was neither alcohol nor marriage but his obstinate belief that quaternions held the key to the universe."[10] Michael J. Crowe, author of a more recent *History of Vector Analysis*, thought that "There is certainly something tragic in the thought of the brilliant Hamilton devoting the last twenty-two years of his life to quaternions, which are *now* of little interest."[11]

Why on earth should anyone think this? The answer seems to lie in sociology, politics and human nature rather than in science. Despite their extraordinary successes, quaternions had a few things which seemed at the time to count against them. Most people wanted a mathematical representation of 3-dimensional space and other vector quantities; they didn't want a fourth term complicating things. The fact that quaternions seemed to *explain* 3-dimensionality rather than just representing it didn't register strongly on their more utilitarian scale of significance. They didn't like the fact that the signs seemed to be all wrong, negative rather than positive for squared quantities. They didn't like the fact that multiplying two quaternions gave a complicated product that had both scalar and vector parts.

Late in the century, Willard Gibbs and Oliver Heaviside set about creating a new mathematical structure out of the wreckage of the old. They called this vector algebra, and it has been the tool used by physicists ever since. They extracted the imaginary part, the 3-dimensional quantity which Hamilton himself had called the 'vector' part from the 4-dimensional quaternion object, made it real, and using Hamilton's original *i*, *j*, *k* notation for this new purpose, called it a vector. Using Hamilton's own terms for the two parts of his quaternion product, they then defined two *separate* products for two vectors, neither of which was a 'product' in the ordinary algebraic sense, and neither of which had any fundamental justification except convenience — a fact which has long been a source of confusion for students of physics. They decreed that to do 'vector algebra' one simply followed a set of rules which had no justification other than the fact that they were convenient. They created an efficient calculating tool for physics as it was then known, but pushed physics in the direction of calculation rather than explanation. It was a remarkable achievement but it lost sight of the whole reason why Hamilton had spent so long on developing the quaternion system.

Two such antagonistic structures could not both survive. To make their victory more secure, the vector theorists waged a ferocious propaganda campaign against the quaternionists. It didn't help that Tait, the main proponent of quaternions, was an older man and not particularly popular, and that the creator of the original system had long been dead, and had produced books that in no sense enhanced anyone's understanding of the subject. The campaign mounted by the vector theorists was very efficient, and after Tait died in 1901, there were no significant supporters of quaternions left. It wasn't the first time that scientists had waged a campaign that stopped at nothing to defeat its rivals, and it surely won't be the last. Scientific battles, unfortunately, aren't always decided purely on the science. At any rate, the quaternionists were totally annihilated and discredited.

What followed was a quite remarkable example of the effects of factionalism and propaganda. The 'uselessness' of quaternions became a legend passed from generation to generation to become a self-fulfilling prophecy. Attempts at using them in serious scientific publications were suppressed with a clinical ruthlessness often by people who didn't know what they were — I think that a good deal less than ten percent of physicists I have come across in my career have any knowledge of quaternions and some haven't even come across the term. Even at the Hamilton bicentenary in 2005, there were well-known physicists claiming that quaternions had never found any use of their own, and only had relevance in being the parent of vector theory. The general impression given — and this is still a long way from losing its force — is that they were the classic example of cleverness for its own sake and didn't materially enhance physical understanding. The fact that this didn't always seem to be the case in practice only served to create an even more bitter opposition in those who had dedicated themselves to their avoidance.

Now, this, by any standards, is bizarre. Quaternions are pure mathematics. Most of pure mathematics was developed long before anyone found a use for it, and many of the uses could not have been anticipated at the time. So it seems incredible that anyone should presume to state that an aspect of pure mathematics must be deemed to be useless *for all future time*, as Bell seems to have done. For some reason, it was considered perfectly acceptable to make statements about quaternions that would have been unthinkable if applied to any other branch of pure mathematics. You can't imagine, for example, anyone defining complex numbers or calculus, or even group theory, as *intrinsically* useless. In any case, quaternions have been used, more or less continuously since their discovery, often by people who would have had no idea that they were actually using quaternions. For example, the axial vectors or pseudovectors, like angular momentum, torque and area, which are used by all physicists, are quaternions pure and simple (as we will show later); the invariant interval used in special relativity $(c^2t^2 - x^2 - y^2 - z^2)$ is actually written in a quaternionic form, in which the scalar quantity (time) takes the real part and the vector quantity (space) the imaginary; the matrices used by Pauli to introduce spin into quantum mechanics are nothing other than a simple development of quaternions which Hamilton had foreseen himself. Apart from creating the whole basis of vector algebra and calculus, and most of its terminology and symbolism, the quaternionists anticipated the space-time connection of relativity and quaternion algebra demanded the introduction of the term that led to spin, long before it emerged in quantum mechanics. When Minkowski produced his 4-dimensional space-time, a quaternionist Theodor Silberstein pointed out that it was essentially quaternionic, but had his (correct) argument summarily rejected.

This is a long way from being just a historical issue. "Those who cannot remember the past are condemned to repeat it."[12] The whole of relativistic quantum mechanics and quantum field theory would have been constructed more easily and more accurately if quaternions had been used instead of the hopelessly cumbersome matrices that are still used as the principal method in most text-books, and many supposedly 'insoluble' problems would now be regarded as solved. Software engineers, who don't have the scruples of twentieth- or twenty-first-century physicists and are driven by commercial demands rather than long-dead academic battles, know that to program for movement of objects in 3-dimensional space, for example in computer games and animations, you have to use quaternions. The same applies to those needing to do great circle navigation, where safety is a major issue.[13] Hamilton's way of thinking was geared to picking long-term winners, rather than immediate impact. This is what happened over his dynamics, which took a century to find its niche in Schrödinger's quantum mechanics. According to Thomas L. Hankins, his biographer: "Over the long term the success of Hamilton's work has justified his efforts. The high degree of abstraction and generality that made his papers so difficult to read has also made them stand the test of time, while more specialized researches with greater immediate utility have been superseded."[14] Towards the end of his life, Hamilton wrote, in a letter to Tait: "Could anything be simpler? Don't you feel as well as think, that we are on a right track, and shall be thanked hereafter? Never mind when."[15] Ultimately, it was Bell, the mediocre historian, who was 'hopelessly wrong', not Hamilton, the 'great mathematician', and opinions like his have had an incalculably damaging effect on physics and its development. A whole century of physicists managed to develop their subject without using the mathematics that gave the only known explanation for 3-dimensionality, but they missed an open goal.

4.4 Vectors

The remarkable thing is that Hamilton had already solved the main problem that quaternions had posed: the fact that it gave results of the wrong sign when the units were squared. Clearly the quaternions, with units 1, i, j, k, were a different algebra from complex numbers, with units 1, i. Though i and i, j and k were all square roots of -1, they were different in a fundamental way: i was commutative — it could be multiplied or multiply anything with exactly the same result; i, j and k were not. Since both algebras were now in existence, it made sense to Hamilton to try to put them together and create a 'complexified' quaternion (or, as Hamilton called it, a biquaternion) algebra. The commutative nature of i makes this very easy to do. Where real algebra has + and − versions of a single unit (1), complex algebra has + and − versions of the units (1, i), and quaternions have + and − versions of the units (1, i, j, k), complexified quaternions have + and − versions of the units

$$1 \quad i \quad j \quad k \quad i \quad ii \quad ij \quad ik$$

The terms ii, ij, ik have particularly interesting properties. They are, first of all, square roots of 1, because $(ii)^2 = i^2 i^2 = 1$, *etc.* They are also anticommutative because the commutativity of i has, of course, no effect on the anticommutativity of i, j, k, and they have interesting multiplication rules:

$$(ii)(ii) = (ij)(ij) = (ik)(ik) = 1$$
$$(ii)(ij) = -(ij)(ii) = i(ik)$$
$$(ij)(ik) = -(ik)(ij) = i(ii)$$
$$(ik)(ii) = -(ii)(ik) = i(ij)$$

Multiplying two of these together does not produce the third term, but i times the third term. So they are not a closed algebra in themselves. But they have the properties required of the long-sought vectors of 3-dimensional space. We can use them to define the mutually perpendicular spatial axes and multiply them by real numbers to indicate lengths in the three directions or positions on the axes. We can also use a new notation, which is close to the one now used for unit vectors. Instead of the red bold italic used for quaternions, we can use red bold symbols for these vector units. So

$$ii \rightarrow \mathbf{i}$$
$$ij \rightarrow \mathbf{j}$$
$$ik \rightarrow \mathbf{k}$$

From now on we will use the convention that italicised symbols square to -1, while bold symbols square to 1. So now we can write the multiplication rules

$$\mathbf{ii} = \mathbf{ij} = \mathbf{kk} = 1$$
$$\mathbf{ij} = -\mathbf{ji} = i\mathbf{k}$$
$$\mathbf{jk} = -\mathbf{kj} = i\mathbf{i}$$
$$\mathbf{ki} = -\mathbf{ik} = i\mathbf{j}$$

Now, just as unit vectors are complexified quaternions, that is, quaternions multiplied by i, quaternions are complexified unit vectors, that is, unit vectors multiplied by i. In other words, since we have both $+$ and $-$ values of all quantities and don't need to be concerned in this context about sign changes, $i\mathbf{i}$, $i\mathbf{j}$, $i\mathbf{k}$ have the exact properties of quaternions. Vectors and quaternions are complexified versions of each other. However, vector algebra has twice as many units as quaternion algebra because vector algebra is a complexified algebra (one in which i appears), whereas quaternion algebra is not. This means that vector algebra must contain quaternion algebra as a subset, but quaternion algebra has no vector subset.

William Kingdon Clifford, working in the later nineteenth century, was the first mathematician to see how this algebra applied to 3-dimensional space. The algebra based on $+$ and $-$ versions of the units

$$1 \quad \mathbf{i} \quad \mathbf{j} \quad \mathbf{k} \quad i \quad i\mathbf{i} \quad i\mathbf{j} \quad i\mathbf{k}$$

which exactly parallels the algebra based on $+$ and $-$ versions of the units

$$1 \quad i \quad j \quad k \quad i \quad i i \quad i j \quad i k$$

is now called the Clifford algebra of 3-dimensional space. In this algebra, all the units have a significant physical meaning. Those based on the real numbers (unit 1) are scalars, or quantities with just numerical values and no directional or dimensional information such as energy and mass. Those based on unit i are called pseudoscalars (imaginary scalars) and include volume and as it would seem from relativity, time. Those based on (i, j, k) are the pure vectors like space and force, that is, 3-dimensional quantities with magnitude (numerical value) and direction. Those based on (ii, ij, ik) are called pseudovectors or axial vectors, and include area and angular momentum. These quantities are 3-dimensional, like vectors, but their directional information is not along a line of action, as it is with vectors like force and momentum, but perpendicular to a plane in which they typically rotate. In principle, vectors define a line along which they act, pseudovectors define an axis perpendicular to the plane in which they act. Because area is a pseudovector and volume a pseudoscalar, while length is a pure vector, we see that all the types of quantity in the algebra are needed just to define space itself.

Clifford also showed that the algebra could be infinitely extended by combining multiple sets of vectors or quaternions or both, and these higher Clifford algebras play a particularly significant role in physics. It is largely through them that we can approach the concept of dimensions higher than 3, including the 10 or 11 of string and membrane theory. However, it is important for people to realise that we can't extend *our* 3-dimensional space to 10 dimensions. Though our space could be a component of some more extended 10-dimensional structure — and this is certainly possible using Clifford algebra — the ten dimensions could not all be like the three we already have. Unfortunately, Clifford, like Maxwell, died young (and in the same year, 1879) and the work wasn't taken up again until almost a century later.

Chapter 5

Symmetry and Duality

5.1 What is measurement?

Three things make up physics at the fundamental level: the laws of physics, the structure of matter, and the fundamental parameters through which everything is observed and theorised about. The first two cannot be described except in terms of the third, so, ultimately, the core information in physics must be connected with the fundamental parameters. We have already indicated that these can probably be reduced to just four: space, time, mass and charge, of which the last two are understood principally as the sources of the four known interactions. So 'mass' means relativistic rather than rest mass, and charge is a composite idea, representing the sources of the weak, strong and electric interactions.

Of the four parameters, space is unique in being the only one that we actually observe, and the only one that we can measure in any direct way. However elaborate our measurement system, it always reduces in essence to a pointer of some kind moving across a scale. Space and spatial attributes — shape, colour, sound vibration — are used in the entertainment industry to simulate things that aren't spatial in origin. The whole success of films, sound recordings and holograms, for example, as make-believe is totally dependent on the principle that we can use space to simulate the action of the other parameters. Remarkably, the measurement of space is not only the unique way of making an observation, it is also universal. Any object of any kind immediately sets up a measurement of space; measurement of space is happening at all times in all parts of the universe.

The closest we come to direct observation of another parameter is of time, but this isn't because we can measure it. Our sense of time passing is nothing to do with the way clocks operate. It is based on the concept of 'entropy'. According to the laws of thermodynamics, any event occurring at any time will lead to the creation of greater disorder in the universe than existed before. Experience gives us a strong sense of how this happens. Say if we drop a cup full of tea onto the floor, we expect the cup to smash into pieces and the tea to spill over the floor. We could imagine trying to restore the cup to its former state, whole and with the tea back inside it, but we know that, even if we had the technology to make this happen, it would require enormous effort and create greater chaos elsewhere as a result. So, if the incident is filmed, we know instinctively whether the film is being played in the correct sequence or backwards.

When we come to the 'measurement' of time, we will see that the thing we are really measuring is space. It isn't the usual measurement of space, because it is repetitive. A single section of space is repeatedly traversed, so that we can use the frequency of repetition to set up the idea of a time interval. Traditional time-measuring devices, such as pendulum clocks and watches with a balance spring, rely on *isochronicity*, the idea that the repetition can be assumed to be regular; the same is true of astronomical measures, such as the yearly orbit of the Earth round the sun and the daily rotation of the Earth. More modern devices such as atomic and digital clocks use some kind of internal oscillation, which can then be counted automatically. Sometimes (as in the hour-glass), the space is only traversed once to fix an interval, but is then used as a standard for the next measurement when the process is repeated.

What is clear from all these examples is that time 'measurement' requires special conditions in which we count repetitions of the same spatial interval. It also relies on acceleration and force. Even using a light signal over a known distance has the same character as the use of traditional and modern clocks. Like Lewis Carroll's White King with two messengers, one for going and one for coming back, it cannot be done without a reflection back to the starting point, so requiring both force and repetition. Now acceleration and force are second order in time (rates of change of rates of change) and so are yet a further remove from direct time measurement. The same requirement applies also to mass and charge. The standard method of measuring mass is by using the force of gravity at the Earth's surface, and for astronomical objects, it invariably uses this force. Charge, of course, is not even detectable without force. In these cases, we usually require a spatial measurement *and* the use of a clock, with its spatial repetitions.

It is surely significant that although we seemingly need four parameters, at least, to construct a description of the world, only one of these is susceptible to direct measurement. Some people have tried to devise an observer-centred physics in which there are only observables, but it seems pretty clear from the physics that has been most successful in explaining nature, and from quantum mechanics in particular, that no such observer-centred exclusiveness is actually possible. Measurability and observability are not universal aspects of nature, and there is no reason to suspect that anything else is either. Some people also, beginning with Descartes in the seventeenth century, have tried to build a theory based purely on space or extension, and to some extent, this was also true of the unified field theory sought by Einstein. But if the best theory we have says that an aspect of physics that we think would be particularly desirable does not lead to correct results, then we should stop desiring it. Quantum mechanics seems to be telling us that the attempt to show an identity between physical concepts such as space and time, while having some interesting consequences, will not, ultimately, be successful. That is why, besides space, the only observable and measurable quantity, we also need time, mass and charge.

5.2 Conserved and nonconserved

If we have not so far, despite considerable efforts, managed to reduce all physical concepts to aspects of space, this doesn't mean we should avoid looking for *relationships* between space and other concepts. There is an alternative in physics to establishing an identity between concepts and this is to establish a *symmetry* between them.

In principle, a perfect symmetry between two things would mean exact identity between them in some respects and exact oppositeness in others. To take a simple example, my left and right hands — which are not, of course, perfectly symmetrical, but which, superficially at least, show a high degree of symmetry — could be taken to be 'near identical' in all respects but one, the lateral positioning of the parts in one of the three spatial dimensions, and to be exactly opposite in that one. If two concepts were *symmetrical*, then they would look alike in many respects, and we could be led to believe for a long time that they were the same. Only a more extensive search would then show that they weren't. This may well be where we have reached with space and time. In fact, from an abstract point of view, symmetry is a better bet than identity, because showing that all physical concepts are versions of space would not explain space itself, and wouldn't offer a route to such an explanation, whereas establishing a symmetry might give us a route into the concepts themselves.

The most fundamental laws of physics — those that we are most sure are true — are concerned in some way with conservation and *nonconservation*. We have discussed conservation at several points already, but what exactly is nonconservation? In principle, it is the fundamental property of the quantities that we have described as variable, and it is just as definite a property as conservation. It is all to do with *identity*. The units of conserved quantities have identities which remain with them after any number of space and time variations. We call this *local conservation* because it means that a conserved quantity, say electric charge, can't disappear in one place and re-emerge in another. Nonconserved quantities have *no identity*. Their units cannot be singled out and labelled like those of conserved ones. One unit of a nonconserved quantity is as good as any other. This leads to three major symmetries. The translation symmetry of time says that one moment in time is as good as any other. We can't pin down the moment. 'Translation' in physics means a linear movement, and the translation would be a linear movement along the time direction. The translation symmetry of space says that one element of space is as good as any other. Space, as a 3-dimensional quantity, also has rotation symmetry. This says that one direction in space is as good as any other — we saw an aspect of this in our diagrams for vector addition, which allows us to reconstruct the components of a vector along sets of axes in any direction.

These are well-established properties of space and time, the two fundamentally nonconserved quantities. Conserved quantities, by contrast, do not have translation and rotation symmetries. In fact, they have properties that are exactly opposite. They are translation and rotation *a*symmetric. If each unit is unique, then one unit cannot be replaced by another. So, for the fundamental conserved quantities, we have the translation *a*symmetry of mass and the translation *a*symmetry of charge. These are the precise properties which give the units of mass and charge individual identities. For mass (which here means relativistic mass or energy) we have a continual transformation of form but throughout this the unit, even in a continuous distribution, retains its identity.[a]

The question now remains whether charge, as a 3-component quantity, exhibits any property which might be described as rotation asymmetry. Do the three types of charge 'rotate' into each other? Experimental evidence at the moment suggests very strongly that they do not. The three types of charge seem to be separately conserved. Particle physicists describe composite particles like the proton and neutron, which are made up of three quarks, as 'baryons' (originally meaning 'heavy particles'), which makes baryons the only particles with *net* strong 'charges'. If strong charges are conserved, then there is no particle for a baryon to decay into except another baryon. Experiments have repeatedly shown that baryon number is conserved in particle interactions, despite repeated speculative predictions that the proton could decay in such a way as to violate it. Baryons share with leptons the property of being fermions; these are the only particles with net weak charges. If baryons can't become leptons or leptons become baryons due to strong charge conservation,

---

[a]For the record, it is important to note that the local conservation of charge or 'identity' of its units is not in any way compromised by the fact that the wavefunctions of, say, two electrons are 'indistinguishable'. Indistinguishability, which is concerned with whether we can *measure* or *observe* wavefunctions as distinct from each other, is a property of the nonconservation of spatial position and has nothing to do with charge as the source of a fermion's identity, this also being an unmeasurable property.

and if there is a separate law of weak charge conservation, then lepton number must also be conserved. Again, all experimental evidence so far collected has preserved this conservation law as well.

Another key property of nonconserved quantities, which is perhaps more difficult to understand but helps to fill out the picture, is called *gauge invariance*. 'Gauge' (a term borrowed from railway tracks) means 'size'; invariance implies that something is preserved against any changes in something else. This property (which occurs in both classical and quantum physics) requires the definition of a conservative system, as one in which the conservation laws apply. They always apply absolutely, of course, to the entire universe, but many much smaller physical systems can be defined in which they apply to a good approximation, and physics generally works by identifying and defining such systems. Now, according to the principle of gauge invariance, you can keep a system conservative even if you make arbitrary changes in the spatial coordinates if these changes produce no corresponding changes in the values of the conserved quantities such as charge, energy, momentum and angular momentum.

There is an example of this from everyday life which may be quite familiar to many. Electric potential is a scalar quantity, defined as potential energy per unit charge, which is basically a ratio of charge to distance from the charge; electric currents are driven through circuits by a difference in potential between the positive and negative terminals of a battery. It is what we often call 'voltage' because it is measured in volts. Now, we always measure voltage as a difference between two potentials; there is no fixed zero value. One way of fixing a zero is to say it is the potential from a charge an infinite distance away; in practical terms, this might be the Earth or ground, which acts as a sink or source of electric charge. However, the negative terminal on a car battery is often referred to as the 'Earth', even though it is not connected to Earth. The point is that the zero point is irrelevant to the potential *difference*, which determines the energy transfer. What is happening in this case is that the equations are structured so that the distance from which the charge's position is measured (*i.e.* the coordinate) is irrelevant as long as the charge value is conserved.

In quantum mechanics, gauge invariance refers to the phase part of the quantum mechanical wavefunction. In principle, gauge invariance shows that the absolute values of the nonconserved quantities, space and time, are irrelevant as long as those of the conserved quantities are preserved. What it says here is that you can make arbitrary changes in the space and time values in the phase terms, as long as the values of the conserved terms are unaffected. To a large extent, it means that the absolute value of a phase term is unknowable because it has no effect on the physical outcome. So, physical quantities in these circumstances are divided into those whose absolute values are crucial (the conserved ones) and those whose absolute values are irrelevant (the nonconserved ones). It is more difficult to show, but it is a cardinal principle of particle physics that gauge invariance is *local*, just like the conservation laws. To the property of local conservation, we have here the exactly opposite property in local nonconservation.

Concepts like gauge invariance and translation and rotation symmetries tell us that physics equations must not only be constructed in such a way that space and time are not conserved, but also that these quantities must be *explicitly seen* to have a property exactly opposite to conservation. This is why the laws of physics are written in terms of *differential equations*. These ensure that the conserved quantities — mass and charge, and others derived from them such as energy, momentum and angular momentum — remain unchanged while the nonconserved or variable quantities, space and time, expressed as the differentials, dx, dt, vary absolutely. It is also the ultimate origin of the fact that all possible paths must be summed in the path-integral approach to quantum mechanics. Space and time not only have no fixed values, but they must be seen not to be fixed.

The intrinsic variability or nonconservation of the parameters space and time is responsible for many of the aspects of quantum mechanics that cannot be reconciled with naïve realism. Einstein may argue that "God does not play dice" in the quantum state, but the nature of nonconservation suggests that it is inevitable that he does. If space and time are intrinsically nonconserved quantities, we have to accept that they are not fixed and should be subject to absolute variation. The only check to this is that conservation principles must hold at the same time, so restricting the range of variation when systems interact with each other. When the interactions are multiple and on a massive scale, we can even make a classical "measurement". The Copenhagen interpretation may require the "measuring apparatus" to be an intrinsic aspect of quantum mechanics, but the measurement aspect is not the component that is really necessary. It is rather the application of new conservation conditions (via external potentials) which result from an isolated system interacting with the environment external to it (the 'rest of the universe') and restrict the degree of variability that would otherwise exist. The so-called 'collapse of the wavefunction' is simply the introduction of a degree of decoherence by extending our definition of the quantum system to incorporate some aspects of its environment.

If we take a free electron, there is no restriction on its variable quantities — it can be anywhere at any time — because it isn't subject to any conservation principle related to space or time. Now, if we bring it close to a proton, so forming a hydrogen atom (which has a single proton as nucleus), then there are now conservation of energy and conservation of angular momentum principles connecting proton and electron. The electron is free to be in any position at any time *as long as the energy and angular momentum are conserved*. This means that the position of the electron is not fixed and has no meaning as such, but there is a fixed range of variability determined by the things that must be conserved. When the hydrogen atom joins with another hydrogen atom to become part of a hydrogen molecule, the electron still has no fixed position at any time, but the range of variability is further modified by the new energy and angular momentum conservation conditions produced by the combined molecule. Eventually, the extension of the system becomes such that the variability falls below limits that we can observe.

Now there is a well-known mathematical result Noether's theorem which in effect requires the pairing of conserved and nonconserved quantities, exactly as happens with the parameters. According to this theorem, to every variational (*i.e.* variable) property there is a conserved quantity. Three famous examples of this have long been established.

 translation symmetry of time ≡ conservation of energy

translation symmetry of space ≡ conservation of momentum

 rotation symmetry of space ≡ conservation of angular momentum

If we take the first of these, we can extend our understanding by using the fact that conservation of energy and conservation of mass (in our sense) are linked by Einstein's famous $E = mc^2$. So, in this case, we are saying that the translation symmetry or nonconservation of time is *exactly the same thing* as the conservation of mass. This is exactly what we would expect from the symmetry we have proposed connecting conservation and nonconservation.

Perhaps, then, we should expect a corresponding link between the nonconservation of space and the conservation of charge — in fact, since we are dealing here with dimensional quantities, we should expect two links, one referring to translation and one to rotation. So we could propose:

 translation symmetry of time ≡ conservation of mass

 translation symmetry of space ≡ conservation of value of charge

 rotation symmetry of space ≡ conservation of type of charge

There is already a kind of realisation of the equivalence of the translation symmetry of space (and consequent conservation of momentum) and the conservation of value of charge, for we know, from work done as long ago as 1927 by Fritz London, that the conservation of electric charge is identical to "invariance under transformations of electrostatic potential by a constant representing changes of phase", and that these are of the kind involved in conservation of momentum. Translated into simpler language, this means that electric charge is conserved, while electric potential (essentially the ratio of charge/distance from source of charge) is not. It is basically the same thing as the gauge invariance we have already discussed. Each of the other two charges has an associated potential of the same kind (a Coulomb potential), which relates to the coupling constant, and so to the 'value of charge'. So the result has a more general application than would at first appear. However, the second result seems totally bizarre. How can the conservation of angular momentum be the same thing as the conservation of *type* of charge, *i.e.* the fact that weak, strong and electric charges do not transform into each other? There is, in fact, an extraordinarily simple explanation, but we need to establish a few other things first.

5.3 Real and imaginary

'Real' in the present context means quantities whose units square to 1 (described as norm 1 in mathematics), and 'imaginary' refers to quantities whose units square to –1 (described as norm −1). This is regardless of whether they are vector or scalar, commutative or anticommutative. Squaring is vital in all aspects of physics, whether in vector addition, amplitudes in quantum mechanics, or interactions of masses and charges. So, whether quantities square to positive or negative values is a significant question. Quantum mechanics has, from the beginning, used imaginary numbers. It has also used non-commutative algebras. But even before quantum mechanics, it was obvious from a relatively early date that any mathematics involving waveforms of any kind was greatly simplified if complex numbers were used. This was particularly apparent to the electrical engineers who worked out the theory of alternating current. And since the introduction of Minkowski space-time, physicists have become totally accustomed to using 4-vectors, or quantities with 3 real parts and one imaginary. The classic example is space and time, where, from Pythagoras' theorem in 4 dimensions,

$$r^2 = x^2 + y^2 + z^2 - c^2t^2 = x^2 + y^2 + z^2 + i^2c^2t^2$$

we can extract the 4-vector

$$r = \mathbf{i}x + \mathbf{j}y + \mathbf{k}z + ict.$$

Here, we represent the unit vectors in blue for the reader's convenience, though coloured symbols wouldn't normally be used in books on mathematical physics. At a later stage, it will be convenient to use a second set of symbols in red.

Similarly from the relation between energy and the three components of the vector momentum (with $c = 1$)

$$E^2 - p_x^2 - p_y^2 - p_z^2 = -i^2 E^2 - p_x^2 - p_y^2 - p_z^2$$
$$p_x^2 + p_y^2 + p_z^2 - E^2 = p_x^2 + p_y^2 + p_z^2 + i^2 E^2$$

we can extract a 4-vector

$$\mathbf{i} p_x + \mathbf{j} p_y + \mathbf{k} p_z + i E.$$

With the c terms included, it would be

$$\mathbf{i} p_x c + \mathbf{j} p_y c + \mathbf{k} p_z c + i E.$$

The 4-vector structure has three real components of space and one imaginary component of time, or three real components of momentum and one imaginary component of energy. If the $3 + 1$ representation of space and time is simply a mathematical 'trick', which is how it is occasionally described, we have to explain why a mathematical trick can explain such a profound physical consequence. We have, in addition, several perfectly good reasons for believing that time 'really' is imaginary in the mathematical sense. First of all, physics repeatedly tells us that quantities containing time pure and simple (for example, uniform velocity) have no meaning; only those in which the time is *squared* — such as acceleration and force — are physically significant. Time 'measurement' always requires force and acceleration, even when we use light itself. There is no one-way speed of light; there has to be a reflection. This is exactly what we would expect from an imaginary quantity. Again, imaginary numbers always have dual solutions, $+$ and $-$, which can't be distinguished. So any equation involving i has to have a partner equation involving $-i$. Now, we know perfectly well that time is one way — as we have already seen, we know this from the increasing amount of disorder we detect after any physical event. Nevertheless, physics equations and physical laws are so constructed that we have two directions of *time symmetry*. Physical meaning can be extracted from reversing the sign of the time parameter, even though we can't reverse time. This is known as the reversibility paradox, but it is no paradox if we believe that time is actually mathematically imaginary, as imaginary numbers always have two mathematical signs. A similar thing happens in relativistic quantum mechanics where we have two signs of the energy parameter (which derives from time via $\partial/\partial t$), but only one sign of physical energy.

When we look at mass and charge, we see possible signs of a similar, but mirror-imaged, structure, though the picture would appear, at first sight, to be greatly complicated by the very distinct differences between the four interactions. It is my belief, on such occasions, that strong suggestions of a symmetry mean it is probably there, so you should assume the symmetry is exact in principle, and that the reason for the apparent complications will emerge in the analysis. The subsequent section on the Clifford algebra will show a striking demonstration that this is indeed the correct assumption.

The differences between the weak, strong and electric interactions seem to be a classic example of a *broken* or *hidden* symmetry. Here, a perfect symmetry exists at some level, but not at the one in current focus, perhaps because of competition with another symmetry, and this is what we will propose here and demonstrate in a later section. In the case of the four interactions, it is believed that there is an energy regime at which the weak, strong and electric interactions would shed all differences and become alike. Now, all four interactions do have one component which is similar. This is a 'Coulomb' term, in effect an inverse square force term. It is the only component of gravity and the electric interaction; but the weak and strong interactions also have additional terms, giving them additional properties. At Grand Unification, the high energy regime where the different forces are expected to become equalised, it is likely that these extra components will shrink and the three nongravitational interactions will become purely Coulomb.

An inverse square force or Coulomb term is exactly what we would expect for a point source in a 3-dimensional space with spherical symmetry. In the weak, strong and electric interactions, it is the 'coupling constant', which, in the electric case, is proportional to the quantity of electric charge. In quantum terms, it is the probability of absorbing or emitting a photon, the quantum of the electric interaction; and in the case of the other interactions, it is the probability of emitting or absorbing the relevant boson or carrier of the interaction. Though charge is really a pure number, a measure of the presence or absence of the source of one of the forces, it will often be convenient to refer to the coupling constant as the charge.

Now, if we compare gravitational and electric interactions, and the Coulomb or inverse square components of the other interactions, there is an age-old problem which has never been satisfactorily resolved. Identical masses attract but identical charges of any kind repel. If we write down the force laws for gravity between masses m_1 and m_2 and electric forces between charges e_1 and e_2, the force for gravity will be negative, signifying attraction, but positive for the electric force, signifying repulsion.

$$F = -\text{constant} \times \frac{m_1 m_2}{r^2}$$

$$F = \text{constant} \times \frac{e_1 e_2}{r^2}$$

There will only be an attractive force between e_1 and e_2 if they are of opposite signs. The problem is solved quickly, however, if we make the charges imaginary ie_1 and ie_2. The force laws then take up an identical form.

$$F = -\text{constant} \times \frac{m_1 m_2}{r^2}$$

$$F = -\text{constant} \times \frac{ie_1 ie_2}{r^2}$$

However, as there are three 'charges' with the same property, completely distinct from each other, then we can use the mathematics available to us and make them components of a quaternion, say $\boldsymbol{i}s$, $\boldsymbol{j}e$, $\boldsymbol{k}w$, where s, e and w represent the strong, electric and weak charges. Now, just as time became the imaginary fourth part of a 4-vector, with three real parts, so mass becomes the real fourth part of a quaternion with three imaginary parts.

space	time	charge	mass
$\boldsymbol{i}x$ $\boldsymbol{j}y$ $\boldsymbol{k}z$	it	is je kw	$1m$

Not only would this give quaternions a direct role in nature, as well as an indirect role through vectors, it would mean that they were in some sense 'prior' to vectors, the character of space-time being determined by symmetry with charge-mass. Space would then necessarily require an anticommutative algebra and be restricted to three dimensions. There is also a bonus: the extra condition requiring space-time and charge-mass to be *absolutely* symmetric requires that the vector part has the character of a complexified quaternion or Clifford algebra, and not the restricted status of the Gibbs–Heaviside algebra. In other words, spin (an extra term that arises in the products of vectors defined as parallel to quaternions) is a built-in requirement and not an unexplained additional extra.

Now it might be possible to argue that we could have chosen mass to be imaginary and charge to be real, and people have occasionally thought about representing gravitational mass as an 'imaginary charge'. However, this does not take into account a fundamental aspect of imaginary numbers: the fact that they only exist as dual pairs with $+$ and $-$ signs. Mass, as far as we know, is 'unipolar'. It only has one sign — whether we make it positive or negative is totally arbitrary, but it only comes in one variety. All charges come in both $+$ and $-$ varieties. This is why we have 'antimatter'. To all particles with a charge structure of any kind, there must be a particle with charges with the opposite signs. We call the process of changing particle into antiparticle or *vice versa charge conjugation*. Even particles with no electric charge, like the neutron, have antiparticles. The neutron has strong and weak charges; the antineutron has strong and weak charges of the opposite signs. However, particles with zero charge structure, like the photon, are *their own* antiparticles. In all cases, it is only the charge structures and the spins that are reversed; the masses are always the same.

A second reason why we have to choose mass as the real term and charge as the imaginary is that we have two ways of apprehending mass physically. One is directly, as the pure quantity, through inertia (force = mass × acceleration). The other is as the squared quantity in gravitation. In principle, we can apprehend mass even if no other mass is present. This does not apply to charge, which we can only apprehend as the squared quantity in Coulomb's law. The same distinction applies to space, which we can apprehend directly (through measurement) and as a squared quantity (through vector addition or Pythagoras' theorem), and time, which we can only apprehend as a squared quantity (through force and acceleration). We can only ever apprehend an imaginary quantity like charge if another charge is present. The imaginary status is not a mathematical convention; it expresses a real physical property.

5.4 Commutative and anticommutative

Our previous discussion has established that mass and time, as scalar and pseudoscalar quantities, are commutative and that space, as a vector, is anticommutative. If charge is fundamentally quaternionic, then it will be anticommutative like space. A result of anticommutativity is that the quantity becomes both dimensional and specifically 3-dimensional. In principle, dimensionality is meaningless at this level unless it is accompanied by anticommutativity. So, time and mass are nondimensional or, as it is sometimes described, one-dimensional.

Already, then, we have a strong indication that mass/time and space/charge have diametrically opposite characteristics regarding commutativity and anticommutativity, but there seems to be another very significant consequence. Anticommutativity introduces a concept of discreteness or discontinuity in that the three components of an anticommutative system act like a closed discrete set. We can take this further by postulating that quantities that are dimensional (or anticommutative) are also discrete or divisible, while quantities that are nondimensional (or commutative) are also continuous or indivisible.

It is easy to see why continuous quantities cannot be dimensional; a dimensional system requires an origin, a zero or crossover point, and this is incompatible with continuity. We can also see that a quantity with only one dimension cannot be measured, because the crossover points to another dimension are needed to do the scaling. It is often said that a point in space shows zero dimensions, a line one dimension and an area two dimensions, but a line is not really a one-dimensional structure, because one dimension has no structure. It is a one-dimensional structure that can only exist in a two-dimensional world, and even this can only exist in a three-dimensional space because two dimensions will necessarily always create a third.

So can we say that space in being 3-dimensional is necessarily also discrete? Undoubtedly. Space has to be discrete, because we would otherwise be unable to observe it. Now, we know that *charge*, also postulated as a 3-dimensional parameter, is certainly discrete. It comes in fixed point-like units or 'singularities', and these can be counted. But space is not like this. So how can both charge and space be discrete? The answer is that space's discreteness is different from that of charge because it is a nonconserved quantity and so has no fixed units. Its discreteness has to be endlessly reconstructed. It is *infinitely* divisible.

This brings us to a profound distinction between space and time, in addition to their mathematical distinctness as real and imaginary quantities. Time, as a nondimensional quantity, cannot be discrete and must be continuous. This has many physical consequences. Time, for example, must be irreversible. Reversing time would mean creating a discontinuity or a zero point. Realising this completes the solution of the reversibility paradox. Irreversibility comes from the continuity of time, while the two signs of time symmetry come from its imaginary nature. Again, because time is not discrete, it cannot be observed, as quantum mechanics tells us. Observables must be discrete. Because it cannot be observed, we always treat it as the *independent variable*, by comparison with space. That is, we write dx/dt, not dt/dx, signifying that time varies independently of our measurements, represented by dx, which respond to the variation in time.

It is the absolute continuity of time which means that, unlike space, it is not *infinitely* divisible. In fact, it is not divisible at all. Clocks do not measure time, but a space with which it is only indirectly related; and infinite divisibility and continuity are absolute opposites, as different as any physical or mathematical properties can be. The absolute continuity of time is crucial in explaining one of the oldest paradoxes in physics: Zeno's paradox of Achilles and the tortoise, which is related to his paradox of the flying arrow already mentioned. Here, Achilles is ten times faster than the tortoise, who, in a hundred-metre race, gets a start of ten metres. So Achilles runs the ten metres to catch up and takes one second. Meanwhile the tortoise has run one metre. So Achilles runs this metre while the tortoise runs a tenth of a metre. Then Achilles runs the tenth of a metre while the tortoise runs a hundredth of a metre. Though he is ten times faster, Achilles will never actually catch up.

Many authors have looked at this problem over the centuries and some have seen that it lies in the assumption that one can divide time into observational units like space. The philosopher G. J. Whitrow, for example, writes that: "One can, therefore, conclude that the idea of the infinite divisibility of time must be rejected, or... one must recognize that it is ... a logical fiction."[16] Motion is "impossible if time (and, correlatively, space) is divisible *ad infinitum*". And the science writers Peter Coveney and Roger Highfield propose that: "Either one can seek to deny the notion of 'becoming', in which case time assumes essentially space-like properties; or one must reject the assumption that time, like space, is infinitely divisible into ever smaller portions."[17] However, most of the commentators seem to be reluctant to draw the logical conclusion that the paradox, and several others proposed by Zeno and others, is a result of assuming that space and time are quantities with the same physical properties. In principle, space is 'infinitely divisible into ever smaller portions'; time is not divisible at all. The so-called "divisions of time" are not, in fact, observed through time.

One way of 'solving' the paradox is to recognise that it involves calculus, and that calculus invokes the procedure of finding the 'limit' of a function as it approaches a particular value. The point where Achilles overtakes the tortoise requires the definition of such a limit. However, one thing that should be recognised about calculus is that it is nothing at all to do with the distinction between continuity and discontinuity. It is to do with whether a quantity is variable or conserved. There are, in fact, two versions of calculus, both dating from the seventeenth century, that are, in effect, based on respective differentiations with respect to space and time. Most students learn how to differentiate using *infinitesimals*, that is you have a term y which is some known function of x. You then increase x by a small amount δx, which creates a corresponding increase of δy in y. You then take the ratio $\delta y/\delta x$ and see what happens when you make δy and δx infinitesimally small, cancelling out any terms that have δx in them.

This was the procedure introduced by the earliest pioneers, and made into a general algorithm by Newton and Leibniz. In my student days, it was considered simple and effective, but not a 'correct' way of differentiating. For that you had to learn the procedure of 'taking the limit'. Newton, who was specifically trying to represent changes in motion in space over time and was dissatisfied by the arbitrary nature of infinitesimals, introduced the first version of this procedure, and it was perfected by Cauchy in the nineteenth century. After some vigorous eighteenth century debates about whether calculus was valid at all, in that at some point a nonzero quantity abruptly became zero, it was Cauchy's version that was sold to students as the 'correct' procedure which solved the philosophical dilemma. Of course, physics students, like myself, continued to use infinitesimals, but mathematicians, who were more concerned with mathematical rigour, no doubt used limits. I remember thinking that even the Cauchy method involved a transition which could not be justified on the basis of previous mathematics.

However, from the 1960s, a new development called nonstandard analysis, largely due to Abraham Robinson, showed that the method of infinitesimals was, after all, just as rigorous as standard analysis or the method of limits.[18] Theorems that could be proved in one theory also held up in the other, and sometimes one method would be more efficient in deriving solutions and sometimes the other. It turned out that this development was related to a similar one in the definition of real numbers.

It is very common to represent one dimension in space by a continuous horizontal line, called the real number line, which is precisely the horizontal axis used in the Argand diagram. Now real numbers start with the natural numbers used in counting: $1, 2, 3 \ldots$. The first extension of this is to admit the negative, and then we have what we call the integers: $\ldots -3, -2, -1, 0, 1, 2, 3 \ldots$. Then we have the rational numbers which include the integers as a special case, but also ratios of integers such as 2/3, 7/5, *etc.* However, there are further numbers which occur in algebraic equations, such as $x^2 = 2$, which cannot be expressed by finite ratios, but which still have a place on the real number line. So we know that $\sqrt{2}$ is greater than 1.4 but less than 1.5, and greater than 1.41 but less than 1.42, and so on. Finally, we have numbers that can't be derived from algebraic equations, but which also have a place on the real number line. The most famous of these, π, is greater than 3, but less than 4, greater than 3.1 but less than 3.2, greater than 3.14, but less than 3.15, *etc.* These are called *transcendental numbers*, and far from being rare, they are actually much more plentiful than integers, rational numbers or algebraic numbers. The total set of numbers is called the real numbers, and is mostly transcendental.

Late in the nineteenth century, Georg Cantor asked the question of whether the real numbers were countable, that is could they be put in a 1 : 1 correspondence with the integers with each integer being used to label one of the real numbers? In fact, this is easily done for the rational numbers, and using a special procedure, can also be done for algebraic numbers, but Cantor concluded that it couldn't be done for the real numbers as a whole. In between every two countable numbers, whether integer, rational or algebraic, there were uncountably infinite real numbers. The real number line was absolutely continuous. Cantor even defined different infinities for countable numbers and for real numbers. There were an infinite number of countable numbers and an uncountably infinite number of real numbers.

Cantor's argument is valid but it is not unique. As early as 1934, Thoralf Skolem had shown that there was an equally valid way of defining the real numbers to be algorithmically countable, and so not continuous in Cantor's sense. This nonstandard arithmetic has a direct relationship to Robinson's nonstandard analysis. In fact, there are two systems of algebra, two of geometry and two of calculus, which depend on two different, equally valid definitions of the real numbers, and there is a perfect duality between the two systems. It would seem that this is a classic case of the 'unreasonable effectiveness of physics in mathematics' that parallels the 'unreasonable effectiveness of mathematics in physics'. There are two methods of differentiation because there are two fundamental variables, one of which is continuous (time) and one of which is discrete (space). There are also two sets of real numbers because one parameter (space) supposes an infinitely divisible line and the other (time) a line that is absolutely continuous. The duality is absolute. However, it is interesting that the method of limits is needed to solve Zeno's problem, because this is the one that applies to continuous time.

The duality in the mathematics also applies to physics itself. Space and time are very different physical quantities. As we will see in the next section, their mathematical connection, which is undoubtedly valid, does not suppose that there is a real physical connection as proposed by Minkowski. This is denied by quantum mechanics where time, unlike space, is not an observable, and the reason becomes clear when we make quantum mechanics relativistic. In fact, when we combine space and time in a 4-vector, we are doing something that is mathematically possible, but physically impossible. To find the nearest physical equivalent, we either make time space-like — the discrete solution — or space time-like — the continuous solution. This is the origin of wave–particle duality: the discrete combination gives us particles, the continuous one waves. The mathematical connection suggests two different physical connections. If space-time actually existed, there wouldn't be any wave–particle duality.

The duality even occurs in the two main forms of nonrelativistic quantum mechanics where Heisenberg follows particle solutions and Schrödinger waves, but it is found throughout the whole of physics, classical as well as quantum. No method has so far been found to validate either the discrete or continuous option at the expense of the other, though the attempt has been made many times. Both Heisenberg and Schrödinger are equally valid, and provide the same answers to fundamental questions, but the ideas of one cannot be mixed with the ideas of the other. Nature is dual because space and time are fundamentally different in respect of their discreteness and continuity.

We have discussed space, time and charge in terms of discreteness and continuity, but what about mass? Both symmetry and its nondimensionality insist that it must be continuous. We are used to thinking of discrete rest masses applicable to particles, but really there is no such thing. All masses are dynamic or relativistic because no particle is free of motion and the energy involved in this. Also, even the so-called rest mass of particles like the electron is generated by the subtle quantum mechanical motion known as the *zitterbewegung*. In addition, mass-energy is a continuum which is present at all points in space in several forms: the Higgs field or vacuum (which requires 246 GeV of energy at every point in space), the 2.7 K cosmic microwave background radiation, the zero-point energy, and even ordinary fields. The continuity of mass is the precise reason why it is 'unipolar', that is, has only one sign. There is no zero or crossover point.

Just as it is neither totally conserved nor totally nonconserved, real or imaginary, nature is neither totally continuous nor totally discrete. It doesn't seem possible that we could define discreteness without also describing continuity. To define something, we also have to know what it isn't. The thing that we don't need, and that seemingly has no place in the physics of matter, is *extended* discreteness, though it does seem to apply to space. Continuity has sometimes been claimed to be an 'illusion', but this is really another appeal to naïve realism. 'Illusions' or ideas are an intrinsic part of reality. We can't even have an idea unless it is somehow part of our abstract picture, and the abstract picture is ultimately the only real world we have. While there have been frequent claims that physics would be better if made totally discrete, continuity always seems to force its way in — for example, through the second law of thermodynamics. If we make the system discrete, as Heisenberg did, continuity appears in the measurement in the Heisenberg uncertainty. If we make the system continuous, as was Schrödinger's option, then discreteness appears in the measurement — the collapse of the wavefunction. Just as with the conserved/nonconserved and real/imaginary distinctions, this one seems to be an exact symmetry of absolute opposites.

Chapter 6

The Fundamental Group Structure

6.1 Dualities

Duality is a kind of pairing in which the two things are not alike but totally determine the characteristics of each other. It seems to be a characteristic of physics as it is also of mathematics. Discovering that such dualities determine seemingly rock-solid *physical* characteristics brings us closer to the union between these two subjects that seems the only possible explanation of their 'unreasonable effectiveness' in extending each other. Here, we have three fundamental dualities with a different pairing of parameters for each:

Conserved	**Nonconserved**
Identity	**No identity**
Mass	Space
Charge	Time
Real	**Imaginary**
Norm 1	**Norm −1**
Space	Time
Mass	Charge
Commutative	**Anticommutative**
Nondimensional	**Dimensional**
Continuous	**Discrete**
Time	Space
Mass	Charge

Such dualities are universal in physics and we see their presence everywhere where the factor 2 or $\frac{1}{2}$ appears in a fundamental context. Remarkably, we can often switch the explanation from one duality to another, indicating that the real explanation lies at an even deeper level. A classic case is the $\frac{1}{2}$ spin of the electron which results in a doubling of the observed quantity known as the magnetic moment (which describes how the electron aligns itself in a magnetic field) in comparison with the expected value. The standard derivation comes from the Dirac equation where the anticommutativity of the momentum operator produces a factor 2, which turns into $\frac{1}{2}$ on the other side of the equation; while the first successful explanation of the effect used a relativistic correction, known as the Thomas precession, for the doubling. The factor $\frac{1}{2}$ for spin or 2 for magnetic moment can thus be derived using both quantum theory and relativity. Yet, in principle, it has nothing to do with either because it is really derived from the intrinsic properties of 3-dimensional space; and there is yet another much simpler explanation of the magnetic effect using purely *classical* theory. Here, we use the distinction between two energy equations, one which describes the kinetic energy acquired during changing conditions (classically $\frac{1}{2} mv^2$), say an object escaping from the Earth's gravity, and the other which describes the potential energy (usually mv^2) which is needed to maintain a system in a steady state, say a satellite in a steady orbit. If we use the kinetic energy equation for the moment at which an electron first aligns itself in a magnetic field, then we easily recover the correct factor without invoking either quantum theory or relativity. All the explanations are true, but the real reason is none of them. The principle of duality is more important than any particular application of it.

Explanation for spin $\frac{1}{2}$	Corresponding duality
kinetic energy equation	conserved/nonconserved
Thomas precession (relativity)	real/imaginary
Dirac equation (quantum mechanics)	commutative/anticommutative

It is fairly obvious, from the table of properties at the beginning of this chapter, that there is some symmetry at work between the fundamental parameters. We could, for example, arrange them as follows:

mass	conserved	real	commutative
time	nonconserved	imaginary	commutative
charge	conserved	imaginary	anticommutative
space	nonconserved	real	anticommutative

We could make it even more obvious by using symbols, such as x, y and z, to represent the properties, with negative versions to represent the exactly opposite 'antiproperties':

mass	x	y	z
time	$-x$	$-y$	z
charge	x	$-y$	$-z$
space	$-x$	y	$-z$

One might realise that this is an example of what mathematicians call a *group*. Groups are ubiquitous at the fundamental level in physics and it should not really surprise us to find one here. To define a group, we need a set of elements (finite or infinite) which are connected by four things:

(1) a 'binary operation' or rule by which any element acts on any other (*e.g.* multiplication, addition)
(2) an identity element, one which when acting on or acted on by any other element using the binary operation produces that element (like the number 1 in ordinary multiplication)
(3) an inverse element to every element in the group; an element operated on by its inverse will give the identity
(4) closure; the binary operation between elements only produces other elements in the group.

A simple example is provided by the units of complex numbers: $1, -1, i, -i$. Here, there are four elements, the binary operation is simple multiplication and the identity element is 1. If we represent the binary operation by $*$, we can draw up a simple multiplication table in which elements in each column multiply elements in each row:

*	1	−1	i	$-i$
1	1	−1	i	$-i$
−1	−1	1	$-i$	i
i	i	$-i$	−1	1
$-i$	$-i$	i	1	−1

Here, we see that we have closure because nothing new is introduced by the multiplications; and $1, -1, i, -i$ have the respective inverses, $1, -1, -i, i$, as we can see from the table by finding the terms which multiply to 1.

Now, the group involving space, time, mass and charge is also a group of order 4 and is almost as simple as that of $1, -1, -i, i$. But we have to devise a more complicated binary operation to include all the properties and 'antiproperties'. Here, we say that combination of any single property or antiproperty with itself gives the property; and combination of any single property with its antiproperty gives the antiproperty. So, we have

$$x * x = -x * -x = x$$
$$x * -x = -x * x = -x$$

and similarly for y and z. This gives us the group table:

*	mass	charge	time	space
mass	mass	charge	time	space
charge	charge	mass	space	time
time	time	space	mass	charge
space	space	time	charge	mass

Here, we obviously have closure, mass is the identity element and each element is its own inverse. However, we could equally well have defined space, time or charge as the identity element, by switching round the definitions of properties and antiproperties. So, if we had written the table of properties in the form:

mass	$-x$	y	$-z$
time	x	$-y$	$-z$
charge	$-x$	$-y$	z
space	x	y	z

then space would have become the identity element, and the group table would have become:

$*$	space	time	mass	charge
space	space	time	mass	charge
time	time	space	charge	mass
mass	mass	charge	space	time
charge	charge	mass	time	space

Two things are striking about these tables. One is the interchangeability of the parameters as abstract objects, even though their physical manifestations have such different effects on human perceptions. The other is the fact that the whole suggests an idea that the total conceptual content of physics at this level appears to be *zero*. Of course, we needn't have specified the properties and antiproperties by algebraic terms such as x and $-x$, or even have used the terms property and 'antiproperty', but there does seem to be a sense in which every conceptual property applicable to nature is negated in some way by one that is its exact opposite. There seems every reason to suspect that the symmetry is absolutely exact. The only possible exception would be if we couldn't derive the apparent deviations from perfect symmetry in the case of charge.

This we will show is most definitely not the case, and no exception to this rule of symmetry has ever been found. The condition, if accepted as absolute, can therefore be used to put constraints on physics to derive laws and states of matter, constraints that are even more powerful and general than those generated by the various laws of conservation on their own. In principle, if the parameters are the primary and sole means through which the physical world is apprehended, then the fundamental symmetries that connect them are not only absolutely true but also *absolutely exclusive*. There is no other fundamental source of physical information.

6.2 Visual representations

If the parameters space, time, mass and charge are perfectly symmetrical in a group structure, then we need only to assume the properties of one of them. The others will then emerge automatically like kaleidoscopic images. Just as we can make any parameter the 'identity' element, the choice of the one we begin with is arbitrary, and we can show this using a number of visual representations. The fact that there are three fundamental properties/antiproperties has an exact analogy in the three primary colours red, blue and green, and the three secondary colours cyan, yellow and magenta. So, we could, for example, represent space, time, mass and charge by concentric circles each divided into three sectors (Figure 14). If we choose any primary colour to represent a property, the corresponding secondary colour will then be the antiproperty. The totality in each sector always adds to zero (represented by white). Two examples are shown; each has multiple interpretations.

Another analogy with the properties could use x, y, z to represent axes in 3-dimensional space, with the $+$ and $-$ directions representing property and antiproperty. The four parameters can then be

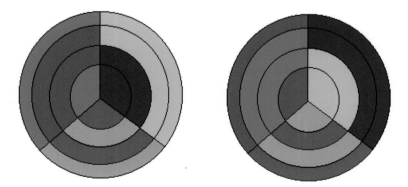

Figure 14. The properties and antiproperties of the four parameters represented by primary and secondary colours.

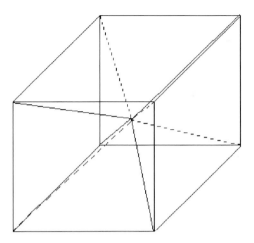

Figure 15. Four parameters, represented by solid lines, and four dual objects, represented by dotted lines.

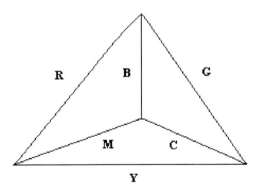

Figure 16. The properties and antiproperties of the four parameters represented by primary and secondary colours (R, G, B and M, C, Y) along the edges of a tetrahedron.

represented by lines drawn from the centre of a cube to four of its corners (Figure 15).

In yet another representation, the parameters could be situated at the vertices of a regular tetrahedron, with the six edges in primary and secondary colours (R, G, B and M, C, Y) to represent the respective properties and antiproperties (Figure 16). Alternatively, we could represent the parameters by the *faces*.

6.3 Two spaces?

A very important aspect of this group structure and the dualities between its components is that the *physical* properties involved can be expressed almost entirely in algebra. The real/imaginary distinction is purely algebraic, as is the dimensional/nondimensional. The conserved/nonconserved distinction seems to be related to the fact that the nonconserved parameters contain the imaginary 'pseudoscalar' i. However, if we see the development of the algebraic structures as a progressive one and related to the construction of a Clifford algebra (as was done in the author's own "universal rewrite system", worked out with colleague Bernard Diaz[1]), then there is good reason to see complex numbers as incomplete quaternion sets and relate this lack of completion to the property of nonconservation. This can also be seen in the reasoning which initially led to quaternions.

We have four separate algebraic systems, of which three are subalgebras (component algebras) of the last.

Mass	Real numbers	1
Time	Imaginary numbers	i, 1
Charge	Quaternions	i, j, k, 1
Space	Vectors	**i**, **j**, **k**, 1, i, i**i**, i**j**, i**k**

Here we use a colour coding that will be important later.

The algebras directly applicable to the parameters also automatically generate their own subalgebras. These include the real number algebra generating the scalar magnitudes of all quantities; and, for space, the pseudovector algebra (for area) and pseudoscalar algebra (for volume). In fact, all the algebras can be seen as subalgebras of the vector or Clifford algebra applied to space. In this algebra, the product of two perpendicular vector units (*e.g.* area) is called a *bivector* and is identical to a quaternion (i**i**, i**j**, i**k** or i, j, k). The product of three perpendicular vector units (*e.g.* volume) is called a *trivector* and is identical to a pseudoscalar or pure imaginary number (i). At that point, with only three perpendicular directions available, the cycle is complete.

We can, in fact, restructure the algebras and subalgebras of space, time, mass and charge using this description:

Space	$\mathbf{i}, \mathbf{j}, \mathbf{k}$	vector		
	$i\mathbf{i}, i\mathbf{j}, i\mathbf{k} \equiv i, j, k$	bivector	pseudovector	quaternion
	i	trivector	pseudoscalar	
	1	scalar		
Charge	$i, j, k \equiv i\mathbf{i}, i\mathbf{j}, i\mathbf{k}$	bivector	pseudovector	quaternion
	1	scalar		
Time	i	trivector	pseudoscalar	
	1	scalar		
Mass	1	scalar		

Significantly, here, the algebras of charge, time and mass *put together* constitute a complete vector algebra equivalent to that of space, because the combination of i and $i\mathbf{i}$, $i\mathbf{j}$, $i\mathbf{k}$ will create the missing $\mathbf{i}, \mathbf{j}, \mathbf{k}$. So charge, time, and mass, *put together*, are equivalent, *mathematically*, to another space. Because it is not a single physical quantity, this composite 'space' cannot be measured in the same way as ordinary space or 'real' space.

In a sense, it is an antispace. It incorporates all the things that *aren't* space, just as all the things that aren't mass combine to an antimass, *etc.* In effect, the symmetry between the parameters is telling us that if these are the only ways that nature can in any sense be apprehended, then the universe is a conceptual 'nothing' or zero with no defining characteristics. It can't be described as either conserved or nonconserved, continuous or discrete, mathematically real or imaginary. We can't even decide whether our view of it is ontological (a "God's eye" view) or epistemological (the view of an observer). And the factor 2 appears everywhere from the fact that everything in physics is really only possible because a mirror image of itself negates its existence.

A zero total *energy* for the universe has long been considered likely. In the early 1960s, Richard Feynman, in a series of lectures on gravitation, observed that the negative gravitational potential energy of the observable universe appeared to cancel out exactly the positive rest energy of the matter composing it ($E = mc^2$): "If now we compare the total gravitational energy $E_g = GM_{\mathrm{tot}}^2/R$ to the total rest energy of the universe, $E_{\mathrm{rest}} = M_{\mathrm{tot}}c^2$, lo and behold, we get the amazing result that $GM_{\mathrm{tot}}^2/R = M_{\mathrm{tot}}c^2$, so that the total energy of the universe is zero. — It is exciting to think that it costs nothing to create a new particle, since we can create it at the center of the universe where it will have a negative gravitational energy equal to $M_{\mathrm{tot}}c^2$. — Why this should be so is one of the great mysteries — and therefore one of the important questions of physics. After all, what would be the use of studying physics if the mysteries were not the most important things to investigate."[19]

Big bang theorists routinely claim that the universe started in 'nothing', that is, zero space, time and matter. Peter Atkins, chemist and science writer, is not untypical in claiming that "the seemingly something is elegantly reorganized nothing, and ... the net content of the universe is ... nothing."[20] Classical physics is strongly based on the idea that the total force in the universe must always be zero — to every action there is an equal and opposite reaction. In one way or another, these ideas seem to stop short of the possibility that there is absolutely *nothing at all* in the 'universe' or in 'nature', even *conceptually*.

At first sight, this might seem startling, because all around us we appear to see 'something', but this is a mistaken way of looking at nothing. We have, in fact, no idea at all what nothing is. It is incomprehensible and impenetrable. From 'inside' nothing, you can't work out what it is like on the 'outside'. We have many equations with zero on the right hand side, but we can't realise any of them in practice. We can write down $2 - 2 = 0$, but we have no idea what this 'really' means. We can get 'close to zero', but if it isn't actually zero, then it isn't really close at all. For example, we can get to within a millionth of a degree above the absolute zero of temperature, but it is still infinitely far away from the zero itself. In fact, absolutely nothing is an ideal starting point for a fundamental theory because it is the one idea we cannot possibly explain. Totality zero would also be an extremely powerful constraint, much more so than, say, conservation of energy, forcing us into a holistic view of the universe such as quantum mechanics seems to require.

6.4 A unified algebra

What happens if we try to put together a unified mathematical structure for the whole system? Here, we need to do some higher algebra, but it involves no calculations and can be thought of as like playing with 'counters'. One way of looking at it is to say we have two vector spaces codified by red and blue symbols. These are independent of each other, or in the usual terminology *commutative*. Each acts as though the other didn't exist. The fundamental units consist of $+$ and $-$ versions of

i	**j**	**k**	i**i**	i**j**	i**k**	i	1
i	j	k	ii	ij	ik	i	1
	vector		*bivector*		*trivector*		*scalar*

The product of each term with every other is called the tensor product, and consists of 64 terms, $+$ and $-$ values of the following:

i	**j**	**k**	i**i**	i**j**	i**k**	i	1
i	j	k	ii	ij	ik		
ii	**ij**	**ik**	i**ii**	i**ij**	i**ik**		
ji	**jj**	**jk**	i**ji**	i**jj**	i**jk**		
ki	**kj**	**kk**	i**ki**	i**kj**	i**kk**		

We could just as easily have begun with the four algebras of space, time, mass and charge:

i j k	i	1	$i\,j\,k$
space	time	mass	charge
vector	*pseudoscalar*	*scalar*	*quaternion*

This would give us

i	**j**	**k**	i**i**	i**j**	i**k**	i	1
i	j	k	$i$$i$	$i$$j$	$i$$k$		
ii	**i**j	**i**k	i**i**i	i**i**j	i**i**k		
ji	**j**j	**j**k	i**j**i	i**j**j	i**j**k		
ki	**k**j	**k**k	i**k**i	i**k**j	i**k**k		

which is exactly the same if we swap ii, ij, ik for i, j, k or i, j, k for ii, ij, ik.

Now, we started with 8 units (16 if we include $+$ and $-$ versions), and these create another 24 (or 48) to make a total of 32 (or 64). It is easy to see that the 64 products form a group and the starting units are what we call *generators* of the group. Specifying them alone is all we need to generate all 64 elements. However, 8 is not the smallest number of generators. If we choose different generators from the group elements, we can reduce the number of generators to a minimum of 5. For example,

$$ik \qquad ii \quad ij \quad ik \qquad j$$

which are all elements of the group, will also generate the entire group if we carry out every possible multiplication. And if we use the 'double space' version, we could use

$$ik \qquad \text{ii} \quad \text{ij} \quad \text{ik} \qquad j$$

Neither of these sets of generators is actually unique. We can create sets of 5 generators in many different ways. We can even split the 64 units into $1, -1, i$ and $-i$, and 12 sets of 5 generators, any of which will produce the entire group.

1	i				-1	$-i$			
ii	ij	ik	ik	j	$-ii$	$-ij$	$-ik$	$-ik$	$-j$
ji	jj	jk	ii	k	$-ji$	$-jj$	$-jk$	$-ii$	$-k$
ki	kj	kk	ij	i	$-ki$	$-kj$	$-kk$	$-ij$	$-i$
iii	iij	iik	ik	j	$-iii$	$-iij$	$-iik$	$-ik$	$-j$
iji	ijj	ijk	ii	k	$-iji$	$-ijj$	$-ijk$	$-ii$	$-k$
iki	ikj	ikk	ij	i	$-iki$	$-ikj$	$-ikk$	$-ij$	$-i$

The important result here is that physics always tends to go for the most minimal representation, and though

$$ik \qquad ii \quad ij \quad ik \qquad j$$

does not appear to be as symmetrical at first sight as

$$\mathbf{i} \quad \mathbf{j} \quad \mathbf{k} \qquad i \quad 1 \qquad i \quad j \quad k$$

yet it contains the same information, and, ultimately, the same symmetries. The only difference is that one of the symmetries (that between i, j and k) is now *broken* or *hidden*. Though we can exchange red and blue, $+$ and $-$, vectors and quaternions, *etc.*, all the sets of five generators have the same *structure*, involving one 3-dimensional symmetry which is preserved and one which is broken. Here, **i**, **j**, **k** are attached to the same unit from the other set, i, so their symmetry is preserved; but i, j and k are each attached to completely different units, so their symmetry is broken.

We have at last found out why the symmetry between the three components of charge and their interactions appears to be broken at the level of observation. If we preserve the symmetry of real space (that of **i**, **j**, **k**), the one we observe, we necessarily have to break the one that is unobservable — that of 'charge' (i, j, k) or the unobservable mathematical 'space' that links charge with mass and time — to produce the simplest possible representation that will generate all others. The 'physical' effect of the mathematical operation is striking. We start with the 8 units needed for the 4 parameters:

i	**i j k**	1	$i\ j\ k$
time	space	mass	charge

To 'compactify' these to the 5 generators we remove the three 'charge' units and attach one to each of the other three parameters:

i	**i j k**	1
k	i	j

The result is that we create 3 new 'composite' parameters, each of which has aspects of time, space or mass, but also some characteristics of charge.

ik	ii ij ik	j
quantised energy	quantised momentum	rest mass
E	p_x p_y p_z	m

If we started only with space, time, mass and charge, this will be the *first appearance* of these quantities in physics, and our analysis suggests that superposition of two sets of parameters with different characters to create generators for the group combining their algebras actually *creates* them. It also simultaneously introduces quantisation and relativity into physics, each of these being effectively the establishment of numerical relations between the units of previously unrelated physical quantities. The new structure is called *phase space*, but it is not independent of either of the 'spaces' that go into its making.

This means that the paired conjugate quantities, time and energy, and space and momentum, are not actually independent, for the set involving energy and momentum is partially created from the more primitive set involving time and space. Fully independent quantities are commutative with each other, but dependent quantities are not. Energy and time are therefore anticommutative at the level of the most fundamental units, as are momentum and space. This is exactly what is expressed in Heisenberg's uncertainty principle: $2 \times$ the product of the fundamental units of the two anticommutative terms produces the most fundamental quantum unit of their combination, $h/2\pi$, the quantum unit of angular momentum.

We have seen how the process affects time, space and mass. But we can see that it also affects charge, for now we create three new 'charge' units taking on the respective characteristics associated with these other parameters.

$i\mathbf{k}$	$i\mathbf{i}\ i\mathbf{j}\ i\mathbf{k}$	j
weak charge	strong charge	electric charge
pseudoscalar	*vector*	*scalar*

In effect, as we will see later, the modification of charge shows the nonlocal or vacuum side of the compactification process, while compactification to energy, momentum and rest mass shows the local. Local and nonlocal are not separate things, however. Neither can be defined without the other. Local interactions can have nonlocal consequences, while nonlocal interactions can have local consequences.

In the Standard Model, there is a famous broken symmetry between the weak, strong and electric interactions. Each of these three interactions responds to a different group structure. The group structures have the following names:

$$
\begin{array}{ll}
\text{weak} & SU(2) \\
\text{strong} & SU(3) \\
\text{electric} & U(1)
\end{array}
$$

The group structures are based on a large amount of experimental work, but have no fundamental explanation in the Standard Model. However, it is not difficult to see how they are generated through the 2-component pseudoscalar ($SU(2)$), 3-component vector ($SU(3)$) and single-component scalar ($U(1)$) nature of the weak, strong and electric charges, though this begins with a nonlocal process.

We can regard the 5 group generators as the most efficient packaging of all the information contained in the group structure of space, time, mass and charge, and codified in their algebraic structures. In principle, we should be able to use it to generate all of the physics which is contained in the interactions between fermions, in particular the Dirac equation and the relativistic quantum mechanics of fermions and bosons. In fact, this emerges in an extraordinarily transparent form in which many developments follow immediately from the algebraic structure.

Chapter 7

The Origin of Quantum Mechanics

7.1 Where does the Dirac equation come from?

The reader may have noticed that we started with 5 units and worked out all possible combinations till we got 64. This is, of course, what happens with the gamma matrices of Dirac. In fact, this is exactly that algebra of gamma matrices, only in a more user-friendly form. Of course, quaternions and Clifford algebras were known to few physicists at the time of the development of quantum mechanics, which explains why they played no significant part in that development. But there is no reason now why we should any longer deprive ourselves of this much more powerful alternative algebra. In fact, it is the only way of getting beyond the idea of using the Dirac equation (still recognised as the most fundamental equation in physics) merely as an impenetrable calculating device.

The units that we have used in the algebra are, of course, arbitrary in terms of size. We can use any scalar or pure number value. Let us therefore use the symbols E, p_x, p_y, p_z, m to represent the scalar values to be used for the units of energy, three directions of momentum and rest mass. Combining these with the algebraic units, we arrive at the expression to represent a fundamental unit combining all the generators of the group:

$$(ik E + i\mathbf{i}p_x + i\mathbf{j}p_y + i\mathbf{k}p_z + jm)$$

Now, this is an expression containing many square roots of -1. To understand its physical meaning, we need to square it. This turns out to be remarkably easy, because anticommutativity means that many of the terms produced in the process (20 out of 25) simply cancel with each other, and we are left only with the squares of the

5 terms in the bracket. So

$$(ikE + ii p_x + ij p_y + ik p_z + jm)(ikE + ii p_x + ij p_y + ik p_z + jm)$$

becomes

$$E^2 - p_x^2 - p_y^2 - p_z^2 - m^2$$

which, according to Einstein's energy equation, is exactly zero. Another way to write this is to first do the 'vector sum' of momentum components $(i p_x + j p_y + k p_z)$ and create a single momentum vector $(\mathbf{p})$:

$$(ikE + i\mathbf{p} + jm)(ikE + i\mathbf{p} + jm) = E^2 - p^2 - m^2 = 0.$$

So $(ikE + ii p_x + ij p_y + ik p_z + jm)$ or $(ikE + i\mathbf{p} + jm)$, the object we have found as a possible basis for including space, time, mass and charge in a single structure, turns out to be a rather strange one: *a square root of zero*! Another could be constructed from the conjugate relation in special relativity between time (t), space $(\mathbf{r})$ and proper time (τ) (again leaving out terms in c):

$$(ikt + i\mathbf{r} + j\tau)(ikt + i\mathbf{r} + j\tau) = t^2 - r^2 - \tau^2 = 0.$$

Such an object is well known to mathematicians. It is called a *nilpotent*, and it is no more strange to have a square root of zero that isn't zero than to have a square root of -1. In fact, every time you draw a right-angled triangle, it incorporates such a structure, and there are an infinite number of different ones. Let us, for example, take the triangle with sides 5, 4 and 3 (Figure 17). Here, we will take a simplified form of the structure and replace the 'momentum' vector

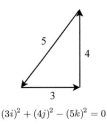

$$(3i)^2 + (4j)^2 - (5k)^2 = 0$$

Figure 17. 3-4-5 Pythagorean triangle represented as a nilpotent.

with a scalar. We can write this in the form $(ik5 + i4 + j3)$. Then, Pythagoras' theorem becomes

$$(ik5 + i4 + j3)(ik5 + i4 + j3) = 5^2 - 4^2 - 3^2 = 0.$$

In effect, going all the way round the triangle either clockwise or anticlockwise takes us back where we started with zero distance travelled. The paths still exist but the result of travelling round them is zero, and we can represent this with $(ik5 + i4 + j3)$ or the vector equivalent $(i\mathbf{k}5 + i\mathbf{4} + \mathbf{j}3)$, where we would find

$$(i\mathbf{k}5 + i\mathbf{4} + \mathbf{j}3)(i\mathbf{k}5 + i\mathbf{4} + \mathbf{j}3) = -5^2 + 4^2 + 3^2 = 0.$$

A more physical example would be the old basic physics experiment in which two cords hanging down from a support and connected at right angles to each other support another carrying a weight which hangs down vertically from the point where they meet. The upward tensions in the two upper cords (T_1 and T_2) are then in equilibrium with the downward tension in the third cord (T_3), which is equal to the attached weight (W) (Figure 18).

Forces really act at the junction of the three cords, but the resultant effect is zero, and we can express this using a nilpotent or square root of zero:

$$(ikW + iT_1 + jT_2)(ikW + iT_1 + jT_2) = W^2 - T_1^2 - T_2^2 = 0.$$

Now, we have already indicated that the algebra we are using has exactly the same form as the gamma algebra used in the Dirac equation of relativistic quantum mechanics, the most important equation in physics. So, if we apply the 'quantisation' procedure to

$$(ikE + ii\mathbf{p}_x + ij\mathbf{p}_y + ik\mathbf{p}_z + jm)(ikE + ii\mathbf{p}_x + ij\mathbf{p}_y + ik\mathbf{p}_z + jm) = 0,$$

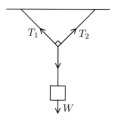

Figure 18. A static system of the tension in two strings supporting a weight represented as a nilpotent.

we might recover a version of this equation, not as an arbitrary assumption or a calculating device, but as an expression of what physics is about at the fundamental level. Using the same 'quantisation' procedure as before ($E \rightarrow i\partial/\partial t$; $p_x \rightarrow -i\partial/\partial x$; $p_y \rightarrow -i\partial/\partial y$; $p_z \rightarrow -i\partial/\partial z$), we obtain an operator of the form

$$(-k\partial/\partial t - ii\mathbf{i}\partial/\partial x - ii\mathbf{j}\partial/\partial y - ii\mathbf{k}\partial/\partial z + jm).$$

This will act on a phase factor, a mathematical object that varies with space and time, to give us back the original expression containing energy and momentum:

$$(ikE + ii\mathbf{i}p_x + ij\mathbf{j}p_y + ik\mathbf{k}p_z + jm).$$

Because we know this will always be the result, we can even use $(ikE + ii\mathbf{i}p_x + ij\mathbf{j}p_y + ik\mathbf{k}p_z + jm)$ to represent the operator.

We have seen previously that the Dirac wavefunction was a spinor with four components, representing fermion/antifermion and spin up/spin down. In this form, this is also true of the operator. Here, it is easy to identify fermion/antifermion as the two signs of E and spin up/down as the two signs of p. So we have:

$$(ikE + ii\mathbf{i}p_x + ij\mathbf{j}p_y + ik\mathbf{k}p_z + jm)$$
$$(ikE - ii\mathbf{i}p_x - ij\mathbf{j}p_y - ik\mathbf{k}p_z + jm)$$
$$(-ikE + ii\mathbf{i}p_x + ij\mathbf{j}p_y + ik\mathbf{k}p_z + jm)$$
$$(-ikE - ii\mathbf{i}p_x - ij\mathbf{j}p_y - ik\mathbf{k}p_z + jm)$$

for the wavefunctions, and

$$(-k\partial/\partial t - ii\mathbf{i}\partial/\partial x - ii\mathbf{j}\partial/\partial y - ii\mathbf{k}\partial/\partial z + jm)$$
$$(-k\partial/\partial t + ii\mathbf{i}\partial/\partial x + ii\mathbf{j}\partial/\partial y + ii\mathbf{k}\partial/\partial z + jm)$$
$$(k\partial/\partial t - ii\mathbf{i}\partial/\partial x - ii\mathbf{j}\partial/\partial y - ii\mathbf{k}\partial/\partial z + jm)$$
$$(k\partial/\partial t + ii\mathbf{i}\partial/\partial x + ii\mathbf{j}\partial/\partial y + ii\mathbf{k}\partial/\partial z + jm)$$

for the operators. It is a key success of the nilpotent representation, not only that these are written down in an explicit form in terms of energy and momentum, or the equivalent operators, rather than merely as unidentified symbols ψ_1, ψ_2, ψ_3, ψ_4, but also that they should have such a direct and symmetric relationship with each other that specifying one immediately specifies all the others by automatic sign changes. This perfect symmetry between the four components is not visible in the matrix formulation of the Dirac equation, which introduces distortions from the structures of the matrices themselves.

In the nilpotent representation there are no unidentified symbols, and the wavefunction is an automatic consequence of specifying an operator, which contains no independent information. There is no black box. The whole calculation is made up of transparent physical information. Also, the spinor structure is an automatic consequence of writing down just the first term in the operator, the other terms acting like 'drones'. Though we should always include all four terms in the operator or in the wavefunction, it will often be convenient to *assume* we mean this when we write down only the first term.

It is not my intention to do calculations here using calculus, but, just to show what one looks like, let's look at the phase factor for a free particle. This would be an exponential term of the form $e^{-i(Et - \mathbf{p} \cdot \mathbf{r})}$, which is a mathematical expression that changes in a known way with changes in the variables time (t) and space ($\mathbf{r}$). This means that the full wavefunction or amplitude ψ would be

$$\psi = (ikE + i\mathbf{i}p_x + i\mathbf{j}p_y + i\mathbf{k}p_z + jm)e^{-i(Et - \mathbf{p} \cdot \mathbf{r})}.$$

Then, operating on the wavefunction with $(-k\partial/\partial t - ii\mathbf{i}\partial/\partial x -ii\mathbf{j}\partial/\partial y - ii\mathbf{k}\partial/\partial z + jm)$ will give us $(ikE + i\mathbf{i}p_x + i\mathbf{j}p_y + i\mathbf{k}p_z +jm)\psi$. So the full operation can be written

$$\begin{aligned}
(-k\partial/\partial t &- ii\mathbf{i}\partial/\partial x - ii\mathbf{j}\partial/\partial y - ii\mathbf{k}\partial/\partial z + jm)\psi \\
&= (-k\partial/\partial t - ii\mathbf{i}\partial/\partial x - ii\mathbf{j}\partial/\partial y - ii\mathbf{k}\partial/\partial z + jm) \\
&\quad (ikE + i\mathbf{i}p_x + i\mathbf{j}p_y + i\mathbf{k}p_z + jm)e^{-i(Et - \mathbf{p} \cdot \mathbf{r})} \\
&= (ikE + i\mathbf{i}p_x + i\mathbf{j}p_y + i\mathbf{k}p_z + jm) \\
&\quad \times (ikE + i\mathbf{i}p_x + i\mathbf{j}p_y + i\mathbf{k}p_z + jm)e^{-i(Et - \mathbf{p} \cdot \mathbf{r})} \\
&= 0,
\end{aligned}$$

which is the Dirac equation for a free particle.

We can regard the four operators in the spinor as *creation operators* for the four possible states:

$$(-k\partial/\partial t - ii\mathbf{i}\partial/\partial x - ii\mathbf{j}\partial/\partial y - ii\mathbf{k}\partial/\partial z + jm)$$

creates fermion spin up

$$(-k\partial/\partial t + ii\mathbf{i}\partial/\partial x + ii\mathbf{j}\partial/\partial y + ii\mathbf{k}\partial/\partial z + jm)$$

creates fermion spin down

$$(k\partial/\partial t - ii\mathbf{i}\partial/\partial x - ii\mathbf{j}\partial/\partial y - ii\mathbf{k}\partial/\partial z + jm)$$

creates antifermion spin down

$$(k\partial/\partial t + ii\mathbf{i}\partial/\partial x + ii\mathbf{j}\partial/\partial y + ii\mathbf{k}\partial/\partial z + jm)$$

creates antifermion spin up

They can be regarded as operators which produce these particular states by acting on vacuum, or the 'rest of the universe', and they are also *annihilation operators* for the opposite states:

$$(-k\partial/\partial t - ii\mathbf{i}\partial/\partial x - ii\mathbf{j}\partial/\partial y - ii\mathbf{k}\partial/\partial z + jm)$$

annihilates antifermion spin down

$$(-k\partial/\partial t + ii\mathbf{i}\partial/\partial x + ii\mathbf{j}\partial/\partial y + ii\mathbf{k}\partial/\partial z + jm)$$

annihilates antifermion spin up

$$(k\partial/\partial t - ii\mathbf{i}\partial/\partial x - ii\mathbf{j}\partial/\partial y - ii\mathbf{k}\partial/\partial z + jm)$$

annihilates fermion spin up

$$(k\partial/\partial t + ii\mathbf{i}\partial/\partial x + ii\mathbf{j}\partial/\partial y + ii\mathbf{k}\partial/\partial z + jm)$$

annihilates fermion spin down

This introduction of creation and annihilation operators acting on vacuum highlights a fundamental fact about the nilpotent theory: it is already a full quantum field theory. In particular, there is no need to introduce the complicated mathematical procedure known as second quantisation, which is normally used to convert quantum mechanics to quantum field theory. In addition, the creation and annihilation operators are completely specified in the nilpotent theory, where in standard quantum field theory they are merely symbols without known structure.

Now, in every case, the operator operates on a phase factor, and this generates an amplitude which then squares to zero. In fact, once the operator is specified, the phase factor is automatically chosen as the only one which will produce an amplitude that squares to zero. The phase factor and amplitude represent, to some degree, the local and nonlocal aspects of the fermion. In principle, the phase factor of the wavefunction is spread through space and time and represents nonlocality, while the amplitude localises a fermion at a point.

We have only chosen one operator — that for a free particle — but we could equally well have written down an operator for a particle interacting with any others by any of the known interactions. We would do this by changing $\partial/\partial t$, $\partial/\partial x$, and so on to covariant derivatives, that is, by adding potential energy terms with structures that describe local interactions, and we could add any number of these terms. In fact, all subsequent expressions for operators can be understood as having these terms. Of course, adding potential energy terms will completely change the structure of the phase factor, which will be completely different from that of a free particle, and the trick is to write down the operator, and then find the only structure of phase factor that will produce a nilpotent solution, that is one that squares to zero:

$$(\text{operator acting on phase factor})^2 = \text{amplitude}^2 = 0.$$

Again, the amplitude will be quite different from the free particle amplitude, $(ikE + i\mathbf{i}p_x + i\mathbf{j}p_y + i\mathbf{k}p_z + jm)$, but it will have the same structure fixed by the algebraic units. That is, there will be a term in ik, ones in $i\mathbf{i}$, $i\mathbf{j}$, and $i\mathbf{k}$, and, finally, one in j. So, we can, for convenience, use $(ikE + i\mathbf{i}p_x + i\mathbf{j}p_y + i\mathbf{k}p_z + jm)$ to represent it simply by redefining E, p_x, p_y, p_z, and m.

The remarkable thing is that to do quantum mechanics, we *don't need an equation at all*. All we need is to define an operator and that will automatically create the phase factor it needs to act upon, eventually ending up with an amplitude that squares to zero. The fermion is, in effect, defined by a set of space and time variations, that is all the places it can be in at the given times. These variations are explicit in the phase factor — the term used for the free particle is equivalent to giving it unlimited range. However, they are *codified* in the operator which is written in terms of differentials in space and time, and finding the phase factor is like unlocking this code. The amplitude which results then expresses the same information using the conjugate parameters momentum and energy.

The fact that we don't need an equation or even a wavefunction to set up a fully relativistic version of quantum mechanics means that we can envisage quantum mechanics not as some separate mathematical structure which we then try to apply to reality, but as something that 'grows' naturally out of the definitions of the fundamental parameters space, time, mass and charge. The operator $(-k\partial/\partial t - ii\mathbf{i}\partial/\partial x - ii\mathbf{j}\partial/\partial y - ii\mathbf{k}\partial/\partial z + jm)$, which contains all the known information about the fermion state and all its interactions and which automatically sets up the corresponding wavefunction, is really a kind of direct expression of the variability of space and time coordinates against the conservation of virtually everything else, exactly as required by the symmetries defined for the fundamental parameters.

7.2 What does the nilpotent wavefunction tell us?

Once we have set up the nilpotent fermion wavefunction, physics results emerge in great profusion simply on the basis of symmetry. We can tell a great deal about the behaviour and structure of fermions and bosons just from the structure of the wavefunction without doing any serious calculations. By making this statement, we are saying that we can derive a great deal of the physics of fermions and bosons simply from the combination of the parameters they represent. Here, it will often be more convenient to use the shortened version of the fermion representation: $(ikE + i\mathbf{p} + jm)$ for $(ikE + ii\mathbf{p}_x + ij\mathbf{p}_y + ik\mathbf{p}_z + jm)$.

We can begin by establishing the distinction between the local and the nonlocal. In effect, everything inside the bracket $(ikE + i\mathbf{p} + jm) = (ikE + ii\mathbf{p}_x + ij\mathbf{p}_y + ik\mathbf{p}_z + jm)$, or equivalently $(-k\partial/\partial t - ii\mathbf{i}\partial/\partial x - ii\mathbf{j}\partial/\partial y - ii\mathbf{k}\partial/\partial z + jm)$, is local and directly concerned with the relationship between E and $\mathbf{p}$ which gives us c; everything outside it is nonlocal. Additions of brackets or switching between them is superposition. Multiplying brackets is combination.

Both of these are nonlocal; they are nothing to do with the relationship between E and $\mathbf{p}$ and the limiting velocity of light. Of course, the nonlocal may have local consequences and *vice versa*. Something outside the bracket, for example another fermion, can affect what is inside the bracket, say, by adding potential energy terms to the energy and momentum operators and making them into covariant derivatives. Once inside the bracket, this information becomes local. If this happens we have reconfigured the system, decohered it, or 'taken a measurement' in the Copenhagen interpretation.

A typical local transition would be to bring one fermion within the range of influence (or field) of another. If all fermions are point particles and their influence is spherically symmetric, then it is convenient to change the coordinates from Cartesian or rectangular coordinates (x, y, z) to polar coordinates (defined by radius r, with one of the point particles at the centre, and two angles). Because a point particle is at the centre of the coordinate system, this is also a centre of physical influence. Dirac worked out a prescription for converting an operator to polar coordinates, replacing the terms in $\partial/\partial x$, $\partial/\partial y$, $\partial/\partial z$, with one in $\partial/\partial r$, but this involved extra terms varying with $1/r$ to the $\boldsymbol{i}$ part of the operator.[21] This has the interesting effect that no nilpotent solution can be found unless a potential energy term varying with $1/r$ is also added to the $\boldsymbol{k}$ part of the operator. This is the characteristic potential energy associated with an inverse-square force, and means, in effect, that if we define a force of any type as emerging with spherical symmetry from a point-particle source, then it has to contain a 'Coulomb' or inverse-square component. This is certainly a feature of all the four known interactions and can be related to the coupling constant. It is completely consistent with the idea that inverse-square forces result from spherical symmetry.

Now, in the nilpotent fermion wavefunction, we have four terms in a superposition, and they represent

$$
\begin{array}{ll}
(ikE + i\mathbf{p} + jm) & \text{fermion spin up} \\
(ikE - i\mathbf{p} + jm) & \text{fermion spin down} \\
(-ikE + i\mathbf{p} + jm) & \text{antifermion spin down} \\
(-ikE - i\mathbf{p} + jm) & \text{antifermion spin up}
\end{array}
$$

The particle we observe would be a fermion, with spin up, the first term represented. But, if we arranged the terms in the order

$$(-ikE + i\mathbf{p} + jm) \quad \text{antifermion spin down}$$
$$(-ikE - i\mathbf{p} + jm) \quad \text{antifermion spin up}$$
$$(ikE + i\mathbf{p} + jm) \quad \text{fermion spin up}$$
$$(ikE - i\mathbf{p} + jm) \quad \text{fermion spin down}$$

the particle observed would be an antifermion, with spin down. If we consider the four terms for the fermion, we can see that the second, third and fourth terms are the three variations that the first term could possibly transform into. We can make these transformations mathematically by multiplying the bracket on either side by i, i; then by k, k; and, finally, by $-j, j$.

$$(ikE + i\mathbf{p} + jm) \qquad \rightarrow \quad (ikE + i\mathbf{p} + jm)$$
$$i(ikE + i\mathbf{p} + jm)i \quad \rightarrow \quad (ikE - i\mathbf{p} + jm) \qquad P$$
$$k(ikE + i\mathbf{p} + jm)k \quad \rightarrow \quad (-ikE + i\mathbf{p} + jm) \qquad T$$
$$-j(ikE + i\mathbf{p} + jm)j \quad \rightarrow \quad (-ikE - i\mathbf{p} + jm) \qquad C$$

These mathematical transformations are also linked with physical ones. The second term has undergone a parity transformation (P), which means that all the spatial coordinates have been reversed in sign, effectively mirror imaged. The third term has had a time reversal transformation (T), which means that the time coordinate has been reversed. The fourth has undergone charge conjugation (C), which is equivalent to parity and time reversal transformations combined, but this is also equivalent to particle being transformed into antiparticle by reversing all the signs of the charges as well as the spin.

It is easy to show that applying two of these symmetry transformations in any order results in the third:

$$CP = T: \quad -j(i(ikE + i\mathbf{p} + jm)i)j = k(ikE + i\mathbf{p} + jm)k$$
$$\rightarrow (-ikE + i\mathbf{p} + jm)$$
$$PT = C: \quad i(k(ikE + i\mathbf{p} + jm)k)i = -j(ikE + i\mathbf{p} + jm)j$$
$$\rightarrow (-ikE - i\mathbf{p} + jm)$$

$$TC = P: \quad k(-j(ikE + i\mathbf{p} + jm)j)k = i(ikE + i\mathbf{p} + jm)i$$
$$\rightarrow (ikE - i\mathbf{p} + jm)$$

while applying all three, in any order, brings us back to the starting point.

$TCP = CPT = \cdots = $ identity:

$$k(-j(i(ikE + i\mathbf{p} + jm)i)j)k = -j(i(k(ikE + i\mathbf{p} + jm)k)i)j$$
$$= (ikE + i\mathbf{p} + jm)$$

This is a well-established result in physics, the TPC or CPT theorem, which states that the laws of physics remain unchanged under the *simultaneous* reversal of space coordinates, time direction and signs of charges. The CPT theorem has been said to be the result of applying relativity simultaneously with causality. In effect, this is the same as applying the condition of nilpotency. In $(ikE + i\mathbf{p} + jm)$, relativity can be said to be the link between ikE and $i\mathbf{p}$; causality is supplied by the third term, jm. The same applies to the conjugate version of relativity based on time, space and proper time. In the nilpotent $(ikt + i\mathbf{r} + j\tau)$, relativity is provided by the link between time and space through ikt and $i\mathbf{r}$, causality by the proper time term, $j\tau$.

Interestingly, the nilpotent structure tells us that though there is a relationship between space and time, it is not a privileged one. Mass and charge are equally linked with them, and there is an equal relationship between all the parameters. Also, an object like $(ikt + i\mathbf{r} + j\tau)$ is not a direct relation between space $(\mathbf{r})$ and time (it) because these are separated by the quaternion coefficients k and i. It takes a separate 3-dimensionality, that of k, i, j, to extend that of the vector $\mathbf{r}$ to 4 dimensions.

The CPT theorem is a significant confirmation of the fundamental significance of the parameters space, time, mass and charge, for all are included except mass. In fact, the only reason why there is no $MCPT$ theorem is because mass has no sign variation. Significantly, there is also no equivalent to 'rest mass' transformation. 'Rest mass' is a purely passive term, unlike the space and time coordinates, and if the energy and momentum are known, it becomes redundant information. It doesn't occur in the phase factor, and we choose to define charge conjugation without reference to it. In fact it is possible to write down a Dirac operator without a mass term if we replace the space and time derivatives with equivalent objects called commutators.[1] There is then no phase factor, and the operator acts directly on the amplitude. This does contain a mass term, but it is the operator, not the amplitude, which defines the particle state.

Chapter 8

Particles and Interactions

8.1 Bosons and the weak interaction

Bosons are combinations, or equivalent to combinations, of fermion and antifermion wavefunctions. In effect they arise from the fact that in any interaction, one state of a fermion is being transformed into another. So it is as if we create the new fermion state and simultaneously an antifermion state to annihilate the old one. To obtain a spin 1 boson, we combine a fermion and antifermion with spins that add to $\frac{1}{2} + \frac{1}{2} = 1$:

$$(ikE + i\mathbf{p} + jm)(-ikE + i\mathbf{p} + jm)$$
$$(ikE - i\mathbf{p} + jm)(-ikE - i\mathbf{p} + jm)$$
$$(-ikE + i\mathbf{p} + jm)(ikE + i\mathbf{p} + jm)$$
$$(-ikE - i\mathbf{p} + jm)(ikE - i\mathbf{p} + jm)$$

We take the superposition of four combination states, multiplying each row separately and then adding them. The result is a scalar quantity or pure number, exactly as boson wavefunctions are expected to be. Its value is $-8E^2$, but all scalar wavefunctions are assumed to be multiplied by the numbers needed to 'normalise' them to 1. Some of these, for example photons, are massless, and these multiply out to the same value:

$$(ikE + i\mathbf{p})(-ikE + i\mathbf{p})$$
$$(ikE - i\mathbf{p})(-ikE - i\mathbf{p})$$
$$(-ikE + i\mathbf{p})(ikE + i\mathbf{p})$$
$$(-ikE - i\mathbf{p})(ikE - i\mathbf{p})$$

Boson wavefunctions, as scalars, add up just like ordinary numbers. So, unlike fermions, we can stack up any number of them together in the same energy state. This is how we are able to create a strong, coherent beam of photons with a laser in a way that could never happen with fermions.

For a spin 0 boson, we combine a fermion and antifermion with spins that cancel to $\frac{1}{2} - \frac{1}{2} = 0$.

$$(ikE + i\mathbf{p} + jm)(-ikE - i\mathbf{p} + jm)$$
$$(ikE - i\mathbf{p} + jm)(-ikE + i\mathbf{p} + jm)$$
$$(-ikE + i\mathbf{p} + jm)(ikE - i\mathbf{p} + jm)$$
$$(-ikE - i\mathbf{p} + jm)(ikE + i\mathbf{p} + jm)$$

The scalar value of the sum of all the products this time is $-8m^2$, again normalised to 1. This time, a massless version cannot exist, as each of the products and their sum is 0.

$$(ikE + i\mathbf{p})(-ikE - i\mathbf{p}) = 0$$
$$(ikE - i\mathbf{p})(-ikE + i\mathbf{p}) = 0$$
$$(-ikE + i\mathbf{p})(ikE - i\mathbf{p}) = 0$$
$$(-ikE - i\mathbf{p})(ikE + i\mathbf{p}) = 0$$

The bosons that are used to transmit the electric, strong and weak forces are spin 1, but there is another boson, discovered in 2012 nearly fifty years after its prediction, that is thought to give mass to all the fermions (the Higgs boson) and this would be spin 0. This particle has a mass and in the theory it replaces a hypothetical massless spin 0 boson (the Goldstone boson) which should not exist, but which has not so far been excluded on fundamental grounds.

Another way to view the nonexistence of massless spin 0 bosons is to say that they represent a weak combination of fermion and antifermion states that cannot both interact weakly. Whether a fermion or antifermion has right-handed spin or left-handed with respect to the direction of motion is determined by the ratio of the signs attached to E and $\mathbf{p}$. If a particle is massless, its helicity (*i.e.* its handedness) is sharply defined — there is no superposition of helicities. If it had a mass, we could argue that it would travel at less than the speed of light and so could be overtaken, and left-handedness would appear as right-handedness, *etc.* However, weak interactions have the unique property that they are *chiral*, or favour one particular handedness. Only left-handed fermions and right-handed antifermions can interact weakly. Now, if we take a combination state such as $(ikE + i\mathbf{p})(-ikE - i\mathbf{p})$, we can see that it requires fermions and antifermions of the *same*, rather than opposite, helicity, bound together by the weak interaction. This is impossible, and the impossibility of the spin 0 boson state tells us why even without invoking the Dirac equation.

The weak interaction is effectively one in which new particle states are formed and it can, to a large extent, be seen as one in which fermion and antifermion unite to create a boson, or in which a boson disintegrates into fermion and antifermion. In quantum mechanics, a system which creates and annihilates bosonic states is called a harmonic oscillator. The classical analogue is something with a periodic motion, like a spring or pendulum.

The weak interaction is local but it also incorporates a nonlocal process. Interactions between charges can be thought of as beginning with the nonlocal establishment of the equivalence of all states that remain unchanged during that interaction. The weak interaction is built into the structure of the fermion itself, which is a superposition of four states:

$$
\begin{aligned}
&(ikE + i\mathbf{p} + jm) &&\text{fermion spin up}\\
&(ikE - i\mathbf{p} + jm) &&\text{fermion spin down}\\
&(-ikE + i\mathbf{p} + jm) &&\text{antifermion spin down}\\
&(-ikE - i\mathbf{p} + jm) &&\text{antifermion spin up}
\end{aligned}
$$

The paired spin states are a necessary consequence of a particle with a nonzero rest mass. However, the positive and negative energy or fermion and antifermion options are really expressions of the fact that the pseudoscalar term i is ambiguous in its $+$ and $-$ signs, and this is carried over into the description of weak charge in the compactification process, which reduces the original 8 generators

to 5. Fermions are the only particles with nonzero weak charges —
in fact this could be the definition of a fermion. Charge carries the
nonlocal or vacuum aspect of the compactification process, so the
fermion wavefunction must be seen to reflect the fact that it represents, among other things, the properties of weak charge, and it
must do this in a way that is nonlocal. This is why it appears as a
superposition of two imaginary or pseudoscalar energy states (ikE
and $-ikE$), with an automatic nonlocal switching between them.
The symmetry is clearly that of the $SU(2)$ group.

Nonlocal processes also have local manifestations. Here it manifests in the fact that the ambiguity and switching between fermion
and antifermion and positive and negative energy could also be seen
as a switching between positive and negative signs of weak charge;
as the switching is nonlocal and therefore instantaneous, this is the
same thing as saying that the positive and negative weak charges
exist simultaneously as a weak *dipole*. Magnetic dipoles are, of course,
common because magnetic north poles cannot exist separately from
south poles. Dipoles of $+$ and $-$ electric charges are also ubiquitous
in nature and very significant; atoms and molecules are classic examples. In these cases the centres of positive and negative charges shift
position continually, so the dipole oscillates. Now single positive and
negative charges repel charges of their own type but attract opposite ones, but dipoles weakly attract other dipoles, and this is the
main cause of physical cohesion, the force which keeps the molecules
in solids and liquids together. Another type of dipole interaction
is responsible for the hydrogen bonding between the two strands
of DNA.

If we consider weak charges as a kind of 'vacuum' dipole exerting
dipolar forces as well as the forces due to single charges, then we
would expect their force laws with *each other* to include a dipolar
term as well as the usual inverse square law for any point source;
in fact the dipole–dipole interaction varies with the inverse fourth
power of the distance; and, where many are combined, there can
be even more complicated terms. This aspect of the interaction for
the forces between the particles, with strength depending on the
distance between them, is local — the potential energies are added to

the derivatives in the operators inside the nilpotent 'bracket' — and can be seen as the local manifestation of the nonlocal superposition of wavefunctions.

The potential energies to be added in the local manifestation are inversely (for the inverse square law Coulomb force component required by spherical symmetry) and inverse-cubely (for the inverse fourth power force required for the dipole–dipole interaction) proportional to the distance. The solution that results is the harmonic oscillator as expected, and the same solution is found for any potential or sum of potential energies that involve dipole or multipole interaction. With the quantum harmonic oscillator, the possible energies are $\frac{1}{2}h\nu, \frac{3}{2}h\nu, \ldots$, separated evenly by $h\nu$. Solving the nilpotent Dirac equation for this combination of potentials gives the exact energy value for this first term to be correlated with fermion spin, suggesting that the spin acts as a kind of weak dipole moment relating the fermion to vacuum.

8.2 Baryons and the strong interaction

Just as the weak interaction manifests itself through a superposition process, so the strong interaction begins in a combination state. Like the superposition, this is nonlocal. The strong charge, as we have seen, acquires three vector components in the compactification from 8 fundamental units to 5 composite ones, and we can see this manifested in the wavefunctions of the only particles that carry a nonzero strong charge, the baryons, such as neutrons or protons. We could write the first line of this wavefunction as

$$(ikE + i\mathbf{i}p_x + jm)(ikE + i\mathbf{j}p_y + jm)(ikE + i\mathbf{k}p_z + jm)$$

As baryons are fermions, with weak as well as strong charges, the baryon would also be a superposition of four combinations of this kind in a spinor structure, but we can leave out this aspect when considering the strong interaction.

Spin can only be defined in one direction at a time, so we can imagine the wavefunction reducing to, say, something like $(ikE + i\mathbf{i}p_x + jm)(ikE + jm)(ikE + jm)$ which, after normalising, becomes simply $(ikE + i\mathbf{i}p_x + jm)$, or to $(ikE + jm)(ikE + i\mathbf{j}p_y + jm)$

$(ikE + jm)$ which becomes $(ikE - i\mathbf{j}p_y + jm)$, or to $(ikE + jm)$ $(ikE + jm)(ikE + i\mathbf{k}p_z + jm)$ which becomes $(ikE + i\mathbf{k}p_z + jm)$. In fact, to maintain the symmetry between the three directions of momentum, we have to consider six possible outcomes, using both $+$ and $-$ values of momentum terms:

$$(ikE + i\mathbf{i}p_x + jm)(ikE + jm)(ikE + jm) \rightarrow (ikE + i\mathbf{i}p_x + jm)$$
$$(ikE - i\mathbf{i}p_x + jm)(ikE + jm)(ikE + jm) \rightarrow (ikE - i\mathbf{i}p_x + jm)$$
$$(ikE + jm)(ikE + i\mathbf{j}p_y + jm)(ikE + jm) \rightarrow (ikE - i\mathbf{j}p_y + jm)$$
$$(ikE + jm)(ikE - i\mathbf{j}p_y + jm)(ikE + jm) \rightarrow (ikE + i\mathbf{j}p_y + jm)$$
$$(ikE + jm)(ikE + jm)(ikE + i\mathbf{k}p_z + jm) \rightarrow (ikE + i\mathbf{k}p_z + jm)$$
$$(ikE + jm)(ikE + jm)(ikE - i\mathbf{k}p_z + jm) \rightarrow (ikE - i\mathbf{k}p_z + jm)$$

These six 'phases' are all valid at the same time — an expression of the rotation symmetry of space or of gauge invariance. The symmetry of this is recognisably the one described by the group $SU(3)$. It has the same structure as the three symmetric and three antisymmetric 'colour' combinations used as the analogy in the standard representation:

BGR, –BRG, GRB, –RGB, RBG, –GBR,

where BGR, for example, represents the quark 'colour' combination blue–green–red, and there are six such colour permutations, three cyclic and three anticyclic. Very significantly, because we have both positive and negative $\mathbf{p}$ terms for the same E term, the six phases incorporate a (maximal) superposition of left- and right-handed components, and so the mass term m is necessarily nonzero, though the unbroken gauge invariance, in which no momentum direction is preferred, means that the boson mediators must be massless, as well as spin 1.

There is currently a so-called 'mass gap' problem in which this minimal positive mass of the baryons is unexplained while the boson mediators are massless. Here, the reason seems to be transparent. Ultimately, and also crucially, this reason, the co-existence of two directions of spin in a fermion state, is recognisable as the *Higgs mechanism*, the only method hypothesised so far for generating the masses of particles from first principles. Many people have argued that the bulk of the mass of a proton or neutron, being derived from gluon transfer, must be different from that which gives mass directly to the quarks, and that the Higgs mechanism will not deliver it. This we can now see is not the case. The Higgs mechanism is completely compatible with massless gluon transfer; and if the Higgs mechanism is true, then this is an argument for saying that it is the universal origin for particle masses.

We can also see immediately what these boson mediators or 'gluons' look like. There are nine possible transitions: $p_x \rightarrow p_x$; $p_x \rightarrow p_y$; $p_x \rightarrow p_z$; $p_y \rightarrow p_x$; $p_y \rightarrow p_y$; $p_y \rightarrow p_z$; $p_z \rightarrow p_x$; $p_z \rightarrow p_y$; $p_z \rightarrow p_z$, although the rules of group theory reduce the total to eight by replacing the three 'colourless' transitions ($p_x \rightarrow p_x$; $p_y \rightarrow p_y$; $p_z \rightarrow p_z$) with two combinations of these. A transition could be seen as a switching of momentum components between the quark brackets, or a switching of the quark bracket positions. In either case, the effect is the same and it always leads to a sign reversal in the **p** term. A structure like

$$(ikE + i\mathbf{j}p_y) \qquad (-ikE + i\mathbf{i}p_x)$$
$$(ikE - i\mathbf{j}p_y) \qquad (-ikE - i\mathbf{i}p_x)$$
$$(-ikE + i\mathbf{j}p_y) \qquad (ikE + i\mathbf{i}p_x)$$
$$(-ikE - i\mathbf{j}p_y) \qquad (ikE - i\mathbf{i}p_x)$$

(if we write it out in full) will replace $-p_x$ with p_y, or $-p_y$ with p_x, because, when normalised, it looks exactly the same as

$$(ikE - i\mathbf{i}p_x) \qquad (-ikE - i\mathbf{j}p_y)$$
$$(ikE + i\mathbf{i}p_x) \qquad (-ikE + i\mathbf{j}p_y)$$
$$(-ikE - i\mathbf{i}p_x) \qquad (ikE - i\mathbf{j}p_y)$$
$$(-ikE + i\mathbf{i}p_x) \qquad (ikE + i\mathbf{j}p_y)$$

The very act of transmitting the strong force by massless spin 1 gluons ensures that the *baryon* has a mass because the change of sign means both spin directions must exist at once.

There are several important consequences of the representation of the baryon wavefunction as a three-component combination state of 'quarks' in which only one component has an angular momentum term and a nonzero i, the presumed carrier of the strong 'charge'. Baryons, such as protons and neutrons, are indeed composed of three 'valence' quarks of the kind included in the wavefunction, but the gluons that carry the strong interaction between them also have strong components of charge (zero in total, because they represent combinations of particle and antiparticle or colour and 'anticolour', but with overall 'directional' aspects). The gluons then split up into virtual 'sea' quarks, producing further gluons, *etc.* The interior of the baryon is effectively a 'strong vacuum' and the mathematical theory needed to describe it, quantum chromodynamics, is very complicated, making the strong interaction the most difficult to pin down in mathematical equations. Much of the angular momentum and mass of a baryon depends not on the valence quarks but on the gluon plasma and the virtual quark sea. The fraction of baryon spin due to the valence quarks, however, can be estimated at 1/3 overall, because only one valence quark is active at any time in contributing to the angular momentum operator, and this is indeed what is found by experiment, the first indication of this surprising result coming from the EMC experiment in 1987. The rest of the spin is then effectively a 'vacuum' contribution, split approximately 3 to 1 in favour of the gluons over the sea quarks, the gluons thus taking half the overall total.

The interaction between the six arrangements of the baryon wavefunction is nonlocal, which means that the rate at which momentum **p** is exchanged between the three components of each arrangement (quarks) is constant, and doesn't depend on the spatial position of the components. There are, however, local consequences. A constant rate of change of momentum with distance is what we call a constant force. A constant force requires an energy which increases with separation (a linear potential). So the energy required to break the quarks apart increases as the separation increases, but decreases as the separation gets smaller, which is the opposite condition to most forces. The large-distance behaviour is called infrared slavery, and the small distance behaviour asymptotic freedom. Clearly the components of each arrangement cannot be parted from each other as they are equivalent to the components of a vector. So we can no more separate the quarks from each other than we can separate the dimensions of space. They only exist as parts of a combined package. This is the origin of infrared slavery. Asymptotic freedom is effectively a statement of the fact that, if the quarks could be imagined in the same position, then no interaction energy would be needed.

A mathematical solution based on a Dirac operator with added Coulomb (inverse linear) and linear potential energies (inside the bracket and so local) gives the expected result, including infrared slavery and asymptotic freedom.[1] This actually completes the nilpotent solutions available from a point source with spherical symmetry of interaction: the Coulomb solution for an inverse linear potential; the quark confinement solution, with infrared slavery and asymptotic freedom, for an inverse linear plus linear potential; and the (weak) harmonic oscillator for the inverse linear plus any other potential. There are no others.

All the interactions participate in the Coulomb solution, but the electric interaction is the only one (except gravity) that is pure Coulomb, the electric charge being the only one not modified by compactification, and therefore leading to no additional superposition or combination states in the wavefunction. The Coulomb solution has been known since the early days of quantum theory. It is also known as the 'hydrogen atom' solution because the hydrogen atom, consisting of one proton and one electron, has exactly the required structure of two point sources of electric charge if we approximate the proton to a point source. The nilpotent procedure is extraordinarily efficient in providing the final result in only six lines of calculation, and it is unique in providing fully analytic solutions of the other two possible conditions relating to a point source of charge with spherical symmetry.[1]

People who learn quantum mechanics using the Schrödinger approach often wrongly assume it must be easier because it is non-relativistic. But this actually makes it more difficult to make sense of. Surprisingly, the more accurate theory is also easier because we can put all the quantities in the 'right place' in the equations. Relativistic quantum mechanics, as it is currently presented using the gamma algebra, is "a mystery inside a riddle inside an enigma"; but using the nilpotent Clifford algebra introduces a massive extension of clarity.

8.3 Angular momentum and charge

We can now consider the special result that we guessed purely from symmetry and that previously looked so inexplicable: the fact that rotation symmetry of space, the conservation of angular momentum, and the conservation of type of charge are the same principle. To explain this, we have to understand that angular momentum conservation is, in fact, *three separate conservation laws* which are completely independent but all required at the same time. For angular momentum to be conserved, we have to conserve separately the magnitude, the direction, and the handedness (*i.e.* whether the rotation is right- or left-handed). The symmetries we require for these conservation laws are the $U(1)$, $SU(3)$ and $SU(2)$ symmetries involved with the electric, strong and weak charges. In effect, these symmetries are versions of the spherical symmetry of 3-dimensional space around a point charge. They say that spherical symmetry is preserved by a rotating system

whatever the length of the radius vector	$U(1)$;
whatever system of axes we choose	$SU(3)$;
and whether we choose to rotate the system left- or right-handed	$SU(2)$.

Conservation of charge is the same thing as the conservation of spherical symmetry for a point source, and it has to preserve three aspects. The $SU(3)$ and $SU(2)$ aspects are dealt with by the respective strong and weak charges, with their vector and pseudoscalar characteristics. All three charges contribute to the $U(1)$ symmetry (just as they do to the Coulomb interaction) because all three charges also have scalar characteristics, but the electric charge is unique in only contributing to this symmetry. So all three charges have to be conserved independently of each other, in the same way as the direction, handedness and magnitude of the angular momentum. It is one of the strongest possible tests of a theory to predict such a totally unexpected result and then to find a simple reason why it must be valid.

8.4 Zero totality

Pauli exclusion tells us that no two fermions can be in the same quantum state. If wavefunctions are nilpotent, this is easily explained, for then the combination state of the two fermions, the product of the two wavefunctions, will be zero:

$$(ikE + i\mathbf{p} + jm)(ikE + i\mathbf{p} + jm) = 0.$$

Another way of looking at this is to suppose that the universe really has a zero totality. Then, if we imagine creating a fermion out of *absolutely nothing*, the *rest of the universe*, from this point of view, is $-(ikE + i\mathbf{p} + jm)$. It is like the 'hole in nothing' required to create the fermion state. It is in every way its mirror image, the total opposite of everything the fermion represents. If the fermion is created at a point (or *creates* a point), this is *everywhere else*. It is what we call *vacuum*. The *superposition* of fermion and vacuum is zero:

$$(ikE + i\mathbf{p} + jm) - (ikE + i\mathbf{p} + jm) = 0$$

and the combination state is zero:

$$-(ikE + i\mathbf{p} + jm)(ikE + i\mathbf{p} + jm) = 0.$$

Vacuum, the dual to the fermion, is effectively the fermion 'inside out', or everything outside the point that the fermion represents. Pauli's exclusion principle now tells us that no fermion can have the same vacuum as any other fermion. No two points can share the same outside. Vacuum, which becomes something of an abstract condition (like conservation of energy) rather than a tangible 'thing', is diffused through all space — it is *nonlocal* — and here we see how it is possible to have instant communication between each fermion state and every other. It is accomplished via vacuum.

Everything that relates to the fermion state, including all its local interactions, is included in $(ikE + i\mathbf{p} + jm)$, the local interactions being fixed by potentials within E and $\mathbf{p}$. However, the vacuum outside any fermion, which incorporates the 'rest of the universe', holds in itself the possibility of changing the fermion state. Here, we see the operation of the 'Observer' or 'Measuring Apparatus' in the Copenhagen interpretation. The Copenhagen interpretation splits the universe into quantum system and Observer/Measuring Apparatus, but the nilpotent structure splits it into quantum system and vacuum. Any change in the vacuum will lead to changes in the E and/or $\mathbf{p}$ terms within the nilpotent bracket or superposition of nilpotent brackets. A change in E and/or $\mathbf{p}$ is necessarily a local interaction and irreversible. Any such change will decohere the quantum system, irrespective of whether there is a real 'Observer'. Schrödinger's cat doesn't even need to be an 'Observer' to produce the change; while the closure of a slit in the Young's experiments is a result of change in the vacuum producing local interactions within the quantum system, and doesn't need to be defined as an intervention of a classical Observer.

Nilpotency specifies that no two fermions can have the same values for E, p and m. We could draw these on axes k, i, j, and then specify that every fermion has a unique direction on these axes, as drawn from the origin. So, no fermion can be massless as $E = p$ would then represent all massless fermions by lines pointing in the same direction.

Nilpotency gives us one way of representing Pauli exclusion, but there is another which is found in most textbooks on quantum mechanics. This is the principle of *antisymmetric wavefunctions*. If we have two fermion wavefunctions, ψ_1 and ψ_2, then their full combination state is represented by

$$\psi_1\psi_2 - \psi_2\psi_1$$

and if $\psi_1 = \psi_2$, this, once again, becomes zero. But fermion wavefunctions are antisymmetric (a property closely related to anticommutativity) and

$$\psi_1\psi_2 - \psi_2\psi_1 = -(\psi_2\psi_1 - \psi_1\psi_2).$$

If we switch the fermions round, the combination state reverses sign.

Now, if we represent ψ_1 and ψ_2 by nilpotents $(ik E_1 + i\mathbf{p}_1 + jm_1)$ and $(ik E_2 + i\mathbf{p}_2 + jm_2)$, we quickly discover that $(\psi_1\psi_2 - \psi_2\psi_1)$ is antisymmetric, as required, but we also find that the only thing, after all the cancellations, that prevents it from being zero is a term which is a multiple of the 'cross product' $\mathbf{p}_1 \times \mathbf{p}_2$ (a multiple which emerges in vector theory), and the only thing that prevents this from being zero is if $\mathbf{p}_1$ and $\mathbf{p}_2$ are pointing in different directions — their relative magnitudes are unimportant. If we now plot the three components of the momentum/spin term $\mathbf{p} = \mathbf{i}p_x + \mathbf{j}p_y + \mathbf{k}p_z$ on a set of axes with units $\mathbf{i}$, $\mathbf{j}$, $\mathbf{k}$, which are the axes of real space, then we find, once again, that we have a unique direction for each possible fermion.

The values of E, p and m and the instantaneous direction of the spin vector each *uniquely* determine the entire knowledge we have of any fermion state (including all its interactions with the rest of the universe). The space represented by $\mathbf{i}$, $\mathbf{j}$, $\mathbf{k}$ (real space) and that represented by k, i, j (vacuum space or antispace) contain entirely dual information, though it is differently structured in each. It is even possible to structure this information as *angular momentum*, which E, p and m and $\mathbf{p}$ provide in different ways, the first through the three symmetries $SU(2)$, $SU(3)$ and $U(1)$, and the second directly (Figure 19).

Now, there is a principle which has been proposed on a few occasions and is widely believed to be true, but which has never been conclusively proved. This *holographic principle*, which has been applied to black holes by Gerard 't Hooft and to string theory by Leonard Susskind, states that the only available information about a system comes on the bounding area.[22] The 'area', however, can be represented as a product of two lengths, or of one length and one time converted to length, *ct*. As momentum is conjugate to space, it can even be angular momentum $\mathbf{r} \times \mathbf{p}$, which is a pseudovector like area

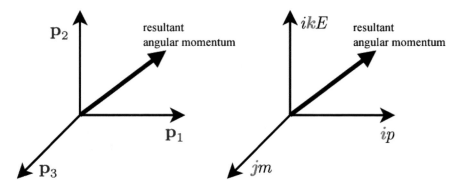

Figure 19. Angular momentum, represented in real space (left) and vacuum space (right).

and which contains the entire information about a system. If we apply the holographic principle to fermions represented by E, p, m, we can see that, because we have a relationship $E^2 = p^2 + m^2$, one of the terms m, which only has one sign, is effectively redundant. The full information comes from just two terms ikE and $\mathbf{p}$, which are conjugate to variables in time and space. It would be logical to extend this to the dual real space $\mathbf{i}, \mathbf{j}, \mathbf{k}$, reducing two of the directions of momentum $\mathbf{p}$ to the conjugate variation in two directions of space.

8.5 Vacuum

The nilpotent fermion is self-dual. Everything that it does in real space is mirrored by the behaviour of its vacuum in vacuum space, or antispace. If we take the four components of the Dirac spinor

$$(ikE + i\mathbf{p} + jm) \quad \text{fermion spin up}$$
$$(ikE - i\mathbf{p} + jm) \quad \text{fermion spin down}$$
$$(-ikE + i\mathbf{p} + jm) \quad \text{antifermion spin down}$$
$$(-ikE - i\mathbf{p} + jm) \quad \text{antifermion spin up}$$

we see that the total energy $(2ikE - 2ikE)$ is zero and the total angular momentum $(2i\mathbf{p} - 2i\mathbf{p})$ is zero. The total charge is zero because the positive and negative energy states also represent fermions and antifermions. If we put it in the operator form, we see that the time and space terms again cancel overall:

$$(-k\partial/\partial t - ii\mathbf{i}\partial/\partial x - ii\mathbf{j}\partial/\partial y - ii\mathbf{k}\partial/\partial z + jm)$$
$$(-k\partial/\partial t + ii\mathbf{i}\partial/\partial x + ii\mathbf{j}\partial/\partial y + ii\mathbf{k}\partial/\partial z + jm)$$
$$(k\partial/\partial t - ii\mathbf{i}\partial/\partial x - ii\mathbf{j}\partial/\partial y - ii\mathbf{k}\partial/\partial z + jm)$$
$$(k\partial/\partial t + ii\mathbf{i}\partial/\partial x + ii\mathbf{j}\partial/\partial y + ii\mathbf{k}\partial/\partial z + jm)$$

The rest mass term is part of the energy, and, as we have seen, can actually be removed entirely from the operator.

The nilpotent structure involves the fermion $(ikE + i\mathbf{p} + jm)$ in interactions with the entire universe. Consequently, its energy is conserved only over the entire universe. It is what we call an open system. There are strictly no conservative (energy-conserving) systems of higher order, though many approximate to this, as Nature tends to reproduce successful structures to the same plan at higher levels. Any interaction with another fermion includes a degree of decoherence, because though the E and m terms are scalars and just add up as numbers, the $\mathbf{p}$ terms, say $\mathbf{p}_1$ and $\mathbf{p}_2$, are vectors with necessarily different directions, and so will always add to a total numerically less than the sum of magnitudes of $\mathbf{p}_1$ and $\mathbf{p}_2$. Energy will always be lost from within the system. Ultimately this manifests itself as the second law of thermodynamics, the law which says that a degree of order is always lost after any interaction and which is the main route through which we perceive the flow of time. Any kind of 'observation', 'collapse of the wavefunction', decoherence or physical change is of this kind, meaning that such processes are necessarily irreversible and dissipative. In quantum physics, the second law tends to manifest itself by the energy possessed by higher-level particle structures gradually degrading over time and passing to lower-level ones such as photons and neutrinos.

If we take an expression like $(ikE + i\mathbf{p} + jm)$ and multiply it by $k(ikE + i\mathbf{p} + jm)$, the result is $(ikE + i\mathbf{p} + jm)$ multiplied by a scalar, which reduces to $(ikE + i\mathbf{p} + jm)$ after normalisation. So multiplying $(ikE + i\mathbf{p} + jm)$ by $k(ikE + i\mathbf{p} + jm)$ has no effect, and we can repeat the process endlessly. So

$$(ikE + i\mathbf{p} + jm)k(ikE + i\mathbf{p} + jm)k(ikE + i\mathbf{p} + jm)\ldots$$

is exactly the same as $(ikE + i\mathbf{p} + jm)$. The term $k(ikE + i\mathbf{p} + jm)$ has the precise characteristics of a *vacuum operator*. The same is true if we multiply the expression by $j(ikE + i\mathbf{p} + jm)$ or by $i(ikE + i\mathbf{p} + jm)$, except that in the latter a vector unit appears in the coefficient, which disappears on alternate multiplications.

$$(ikE + i\mathbf{p} + jm)i(ikE + i\mathbf{p} + jm)i(ikE + i\mathbf{p} + jm)\ldots$$
$$(ikE + i\mathbf{p} + jm)j(ikE + i\mathbf{p} + jm)j(ikE + i\mathbf{p} + jm)\ldots$$

So $k(ikE + i\mathbf{p} + jm)$, $j(ikE + i\mathbf{p} + jm)$ and $i(ikE + i\mathbf{p} + jm)$ are all vacuum terms. But what vacuum are they? We already have $-(ikE + i\mathbf{p} + jm)$ as the fermion vacuum, so how do they connect to this? There is, in fact, another way of interpreting the results. Since $k(ikE + i\mathbf{p} + jm)k = (-ikE + i\mathbf{p} + jm)$, effectively flipping the sign of the energy term and converting the fermion to antifermion, we can rewrite

$$(ikE + i\mathbf{p} + jm)k(ikE + i\mathbf{p} + jm)k(ikE + i\mathbf{p} + jm)k$$
$$(ikE + i\mathbf{p} + jm)k(ikE + i\mathbf{p} + jm)\ldots$$

in the form

$(ikE + i\mathbf{p} + jm)$	$(-ikE + i\mathbf{p} + jm)$	$(ikE + i\mathbf{p} + jm)$
fermion	antifermion	fermion

$(-ikE + i\mathbf{p} + jm)$	$(ikE + i\mathbf{p} + jm)\ldots$
antifermion	fermion$\ldots$

Here, all terms after the first are virtual, a description of vacuum, or the effects of the real fermion represented by the first term on vacuum. We can also recognise that this is what happens in the spinor wavefunction, where the three remaining terms after the first are vacuum reflections of the first in the *three discrete components* of vacuum.

$(ikE + i\mathbf{p} + jm)$	fermion
$(ikE - i\mathbf{p} + jm)$	reflection in strong vacuum
$(-ikE + i\mathbf{p} + jm)$	reflection in weak vacuum
$(-ikE - i\mathbf{p} + jm)$	reflection in electric vacuum

In effect, application of the three quaternion operators k, i and j splits the *continuous* vacuum, represented by $-(ikE + i\mathbf{p} + jm)$, into three components which respond to the strong, weak and electric interactions and carry the nonlocal variations produced by compactification.

In quantum field theory, fermions never exist in a pure and simple state. They are constantly interacting with vacuum, polarising it and producing virtual fermion and antifermion pairs, which, in turn, produce further fermion and antifermion pairs, and so on. This is exactly what we see happening here. When they interact with other fermions, the vacuum interactions change their *effective* couplings. This can be calculated as an *effective* change in the observed value of charge depending on the interaction energy. The process is called *renormalisation*, and gives precise answers for each interaction.

However, there is also a contribution due to each fermion's vacuum interaction with itself or *self-energy*, and this appears to be infinite. Renormalisation can remove this by subtracting one infinity from another, but this has always been considered an unsatisfactory and somewhat arbitrary process. One theory which would make this physically meaningful if it could be experimentally demonstrated is that every type of fermion has a *supersymmetric* boson partner, and *vice versa*. In that case, the summation of terms could be done in such a way that negative fermion energies could be subtracted from positive boson energies to eliminate the infinite term automatically.

No supersymmetric particles have ever been discovered, but our representation of the interaction of fermions with vacuum suggests a way of achieving this objective. Fermion–antifermion combinations are also bosons. So,

$$(ikE + i\mathbf{p} + jm)(-ikE + i\mathbf{p} + jm)(ikE + i\mathbf{p} + jm)$$
$$(-ikE + i\mathbf{p} + jm)(ikE + i\mathbf{p} + jm)\ldots$$

which we have shown is exactly the same as the single fermion state, $(ikE + i\mathbf{p} + jm)$, can be seen as a string of spin 1 bosons, a single boson, or an alternate string of bosons and fermions. In other words, fermions can also be seen as bosons when combined with their own weak vacuum in a way that leaves them unchanged. A fermion produces a boson state by combining with its own vacuum image, and the two states form a supersymmetric partnership. Since the energies are also the same, the supersymmetry between them is also exact, leading to exact cancellation. We have supersymmetry without requiring extra particles.

In standard supersymmetry, we have an operator Q that converts a boson to a fermion and an operator $Q\dagger$ that converts a fermion to a boson. Here, it is clear that Q is represented by $(ikE + i\mathbf{p} + jm)$, or, more fully, by

$$(ikE + i\mathbf{p} + jm)$$
$$(ikE - i\mathbf{p} + jm)$$
$$(-ikE + i\mathbf{p} + jm)$$
$$(-ikE - i\mathbf{p} + jm)$$

and $Q\dagger$ by $(-ikE + i\mathbf{p} + jm)$, or, more fully, by

$$(-ikE + i\mathbf{p} + jm)$$
$$(-ikE - i\mathbf{p} + jm)$$
$$(ikE + i\mathbf{p} + jm)$$
$$(ikE - i\mathbf{p} + jm)$$

A fermion converts to a boson by multiplication by an antifermionic operator (here, it is a vacuum reflection); and a boson converts to a fermion by multiplication by a fermionic operator, and we can represent the sequence by

$$QQ\dagger QQ\dagger QQ\dagger QQ\dagger Q \ldots$$

If we wanted the same for *antifermion* and boson, Q would be $(-ikE + i\mathbf{p} + jm)$ and $Q\dagger$ would become $(ikE + i\mathbf{p} + jm)$, reversing the roles of creation and annihilation.

The universe is overwhelmingly made up of fermions. Antifermions only exist for brief moments when 'pair production' splits up a boson into a fermion/antifermion pair. It has often been postulated that there must at one time have been an equal or near equal number of fermions and antifermions, and that what we have left is the remnant that followed some mutual annihilation. But no mechanism has so far been found for converting matter into matter to explain the 'slight' excess of fermionic matter. Now, Dirac said that fermions have positive energy, antifermions have negative energy. Wheeler and Feynman said that fermions go forward in time, antifermions go backward in time (an idea corroborated by our nilpotent operators where the $\partial/\partial t$ terms reverse sign for antiparticles). We can add that fermions exist in space, antifermions in antispace or vacuum space. Antimatter is almost a *definition* of the vacuum. The switching from fermion to antifermion in *zitterbewegung* is a switching from space to antispace or vacuum space, which is why we never observe the antifermionic options — only the mass produced by the slowing down that the motion causes. If we treat this with the seriousness it deserves, we may well find that searching for the 'missing' antimatter is following a mirage. According to the structure

of the Dirac spinor, there is no excess of matter over antimatter —
there is the same amount of each. It is just that they reside in different
spaces, and only that carrying the matter is observable.

8.6 Electroweak mixing

Returning for a moment to the weak interaction, we see that though
we have already provided the main outline for how it operates,
there are some subtle aspects which it will now be convenient to
consider. For example, if the fermion, switching to antifermion,
already has mixed spin states because of its mass, then the switching
between fermion and antifermion is not only a T or weak transition between positive and negative energy states ($ikE \rightarrow -ikE$ and
$-ikE \rightarrow ikE$) but also a C transition involving both ikE and $i\mathbf{p}$,
with processes such as

$$(ikE + i\mathbf{p} + jm) \quad \rightarrow \quad (-ikE - i\mathbf{p} + jm)$$

$$(ikE - i\mathbf{p} + jm) \quad \rightarrow \quad (-ikE + i\mathbf{p} + jm)$$

$$(-ikE + i\mathbf{p} + jm) \quad \rightarrow \quad (ikE - i\mathbf{p} + jm)$$

$$(-ikE - i\mathbf{p} + jm) \quad \rightarrow \quad (ikE + i\mathbf{p} + jm)$$

The C transition is the one associated with the 'electric vacuum',
the j operator and the rest mass term, suggesting that the weak and
electric interactions become mixed when spins are mixed. If the weak
interaction responds only to left-handed helicity states in fermions,
then right-handed states will be intrinsically passive, except in the
generation of mass, though left- and right-handed states will respond
equally to the electric interaction. Two different $SU(2)$ symmetries
can be identified. The $SU(2)$ of spin is a simple description of the
existence of two helicity states, left- and right-handed. However, there
will also be another $SU(2)$ symmetry, which is called *weak isospin*,
which says that the weak interaction capacity of the fermion is independent of whether or not an electric charge is present and generating
its own contribution to mass. In effect, weak isospin tells us that the

weak and electric interactions, though often occurring simultaneously and in combination, are actually independent processes in origin.

Now, we have said previously that local interactions between charges begin with the nonlocal establishment of the equivalence of all states that remain unchanged during that interaction. In a sense this is what we mean by charge conservation. All other states are equivalent as long as the action of the charge is unaffected. So a weak interaction allows any changes that preserve the value of the weak charge. The same is true of the strong interaction, which is electric charge independent; so protons, which are positively charged, and neutrons, which are electrically neutral, interact, according to the strong interaction, as though they were the same particle. Now, weak isospin allows the transfer of electric charges and mass without affecting the weak interaction, so the interacting weak bosons (W^+, W^-, Z^0), unlike those mediating the other forces, have nonzero masses and can be positively or negatively charged.

If we look again also at the connection between the conservation of angular momentum and the conservation of type of charge, we will see that the strong charge entirely determines the system of axes for $\mathbf{p}$; and the weak charge entirely determines the handedness in the sign of iE (the T transition), except for that part which depends on the mass m, and principally comes from the presence or absence of electric charge (the C transition). So, here, we see directly the distinction between the two versions of $SU(2)$, and the partial connection between them. All the charges contribute to the magnitude, but the electric charge is unique in *only* contributing to this aspect.

Quantum field theory says that the weak interaction responding only to left-handed helicity states in fermions and right-handed helicity states in antifermions requires a 'filled' or 'degenerate' weak vacuum. A degenerate vacuum is one in which there are effectively an infinite number of equivalent states with the same energy and momentum. If we imagine initially a massless state of this vacuum, say ($ikE + i\mathbf{p}$), then it could transform into another state of the form ($ikE + i\mathbf{p}$) via a spin 0 boson through which one of these states is annihilated and the other created. However, a boson of the required form, $(ikE + i\mathbf{p})(-ikE - i\mathbf{p})$, cannot exist as it would be immediately zeroed, which means that the vacuum must be of the form $(ikE + i\mathbf{p} + jm)$ and the boson $(ikE + i\mathbf{p} + jm)$ $(-ikE - i\mathbf{p} + jm)$. This is the principle of the Higgs mechanism, and we can see that it is essentially derived from the principle that massless spin 0 bosons cannot exist because $(ikE + i\mathbf{p})(-ikE - i\mathbf{p}) = 0$.

Chapter 9

Space and Antispace

9.1 Space and antispace

The nilpotent structure suggests that space is not the passive thing we tend to think it is in the context of naïve realism. To see this we can take an example from topology, which is the mathematics used to describe those spatial properties which are preserved when objects are deformed. Suppose that we transport a vector round a circuital path while making sure that its direction is always pointing along a tangent to the path — this is what we call parallel transport. If the space is of the usual kind, we call it 'simply connected'. However, we can imagine a space containing a 'singularity' — a point which makes some kind of break in the space, for example one made by a point particle of matter. This space would be called 'multiply connected'. Now, if we create a circuital path in this space (with the singularity inside the circuit) and parallel transport a vector round this, we will find that when it reaches its starting point again, it will be pointing in the opposite direction. We have to take it *twice* round the circuit to find it pointing the same way. We say that the vector has acquired 180° of phase after one circuit, and needs 360° to end up pointing in the original starting direction. This is called the geometric or Berry phase, and in many of the contexts in which it occurs, it is interpreted geometrically. However, it is likely that its ultimate origin is in topology (Figure 20).

Now, the crucial thing is that the 180° phase is connected with creating a singularity in an otherwise 'normal' space. Pure physical space has no such singularities; however one physical quantity, charge, is constructed of them, and charge is the ultimate origin of

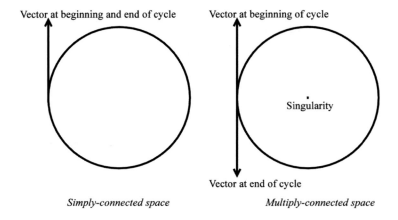

Figure 20. The Berry phase represented topologically.

physical particles. If we take an electron, or any other fermion, as far as we know it has no definable size; and, logically, it seems difficult to find a way in which we could define a concept of *extended* discreteness. But, if an electron is a point, it could be said to exist in *its own* multiply connected space. So, the 180° phase change on rotation, or spin $\frac{1}{2}$, would be intrinsic to it. We already know that real fermions act in exactly this manner. One classic example of geometric phase is the Cooper pairing of superconductivity, where two electrons pair in such a way that they each supply a 180° phase change, or spin $\frac{1}{2}$, to become, in combination, the equivalent of a boson with 360° phase change on rotation, or a total spin 1.

An electron on its own could be said to have the 180° phase or spin $\frac{1}{2}$ because its partner, in this case its mirror-image vacuum state, supplies the rest of the phase or spin. In fact, the representation of an electron wavefunction as two electron plus two antielectron states, and the *zitterbewegung* between them, suggests that the electron spends only half of its time in 'real space' and the rest of its time in 'vacuum space' or 'antispace', and hence only completes half of the circuit when it rotates. Another way of looking at this is to say that it requires a combination of the two 'spaces' to create the multiply connected space that gives us the geometric phase, and that manifests itself as a physical singularity. In this way, we could see the singularity that we call a fermion as being a kind of twisting together of two spaces, each of which sees the other as distorted, with the connection between them secured by the nilpotent structure which ensures that they are dual. So, instead of a storybook picture in which 'tangible matter' finds itself in a spatial complex, we have an abstract entity in which a combination of space and the dual structure which zeros it manifests itself as a space with singularities that provide the points of entry into the dual state. The idea is not totally dissimilar from Roger Penrose's concept of a twistor, which uses a complex 4-dimensional space as a basis to try to derive material particles; but Penrose's idea, coming from a background of general relativity, is based on the reality of 4-dimensional space-time as a *real structure*, which means that particles with nonzero masses have to somehow emerge from the massless photon. In our case, there is no true 4-dimensional structure, only two sets of interlocking 3-dimensionalities. As Einstein wished to show, space does produce its own material structures, but via the inherent symmetry and duality which creates a zero totality.

The nilpotent structure with its origin in a combination of space and a dual antispace suggests that another metaphor for quantum mechanics may be a more productive one than any of those we have so far discussed. This is John Wheeler's "one electron" theory of the universe, now refined to a "one fermion" theory.[23] Here, everything that happens can be reduced to the action of one fermion in a range of backward and forward time states and spatial positions, being equivalent to many fermions acting simultaneously. The metaphor is attractive from the point of view of a fermion in nilpotent theory necessarily defining the rest of the universe as a mirror image of itself, and from the fact that physical events appear to take place in an absolute causal sequence or universal birth-ordering, partly determined by the second law of thermodynamics. Two significant objections to the original version were the apparent imbalance of matter and antimatter in the universe, and the fact that the fermions and antifermions would mutually annihilate. It would seem, however, that both of these objections can now be addressed by the fact that equal numbers of fermions and antifermions exist, but *in different spaces*, so increasing the attractiveness of the one-fermion metaphor.

We could imagine all the possible space and time conditions for the single fermion as constituting the vacuum as 'rest of the universe', and this would include all the states to which the fermion could possibly aspire over time. In this way, a real single fermion would include the entire possible history of the universe within its event horizon, or all possible universes to which it could belong in the many worlds interpretation. (These universes, however, are only vacuum universes, unlike those in the more literal interpretations of the many worlds metaphor.) The interpretation does not support determinism, however, because the entire history could only be defined by localising the fermion exactly, which the uncertainty principle ensures we cannot do. What we have is a precise idea of what we mean by nonlocal, as containing all other potential states, in space and time, with these determined by the real states. To fix a particular moment in time would be 'localising' in time, in the same way as fixing a position is localising in space.

9.2 Can we have 10 dimensions?

One doesn't have to be a dyed-in-the-wool string theorist searching among the 10^{500} possible 'compactifications' from the 10/11 particle dimensions to the 4 observed as space and time to see that the symmetries it produces are important on fundamental grounds. Probably the most famous one comes from the claim that 10 dimensions are the minimum needed to cancel all the anomalies in the theory of fundamental particles. Most people have the idea that this means 9 of space and one of time. However, we can't extend *our* space to 9 dimensions, and string theory actually puts the extra ones below the Planck length — a very small length (1.6×10^{-35} metres) calculated from the fundamental constants G, c and $h/2\pi$, beyond which it is impossible to even contemplate observation. Below the Planck length a 'dimension' cannot be identified as space at all. So the concept of six extra *spatial* dimensions is rather misleading. Such a 'dimension' can be anything you want to call it which gives an extra degree of freedom to a system.

Clifford algebra, with its structure of dimensions on top of dimensions, allows us to look at dimensions in many perspectives, and to view the same structure as having different dimensions in different perspectives. The nilpotent structure could, for example, be seen as 10-dimensional, 8-dimensional, 5-dimensional, 4-dimensional, 3-dimensional or even 2-dimensional, depending on how we define the dimensions. It certainly contains a 10-dimensional structure, which is very much of the kind required by string theory, although the model-dependent aspect provided by the 'strings' isn't needed. Starting with the eight fundamental units

$$i \qquad \mathbf{i\,j\,k} \qquad 1 \qquad i\,j\,k$$

time	space	mass	charge

we have a double modification, even a dual one, when we carry out our 'compactification' to five generators with five 'dimensions' for energy:

$$i k \qquad\qquad i i\;\; i\mathbf{j}\;\; i\mathbf{k} \qquad\qquad j$$

quantised energy	quantised momentum	rest mass
E	$p_x\;\,p_y\;\,p_z$	m

which are conjugate with those for time, space and proper time, and five for charge:

$$i k \qquad\qquad i i\;\; i\mathbf{j}\;\; i\mathbf{k} \qquad\qquad j$$

weak charge	strong charge	electric charge
pseudoscalar	*vector*	*scalar*

As it happens, six of these are conserved quantities for any fermion — that is, all but energy and momentum. We could say, as the string theorists do, that they are *compactified* (in their sense of the term). They are also below the Planck length in that they seemingly have no size at all (which probably means the same thing). I have it on the authority of a colleague that 'The perfect string theory is one in which self-duality in phase space determines vacuum selection'.[24] I am not sure he had in mind exactly what we have here but it certainly fits a description of this kind. The nilpotent wavefunction is written out in phase space, it is self-dual, and it determines vacuum selection. Here, we see that it is the symmetries that matter not the structures.

A more sophisticated concept, membrane theory or M-theory, aims to produce a unification of the different types of string theory so far produced by embedding the 10-dimensional structures in yet another structure or 'dimension'. The nilpotent wavefunctions are, of course, embedded in a Hilbert-type 'space' or algebraic structure which allows us to write down combination states for different wavefunctions, and we could consider that as serving the same purposes as the so-called "eleventh dimension" of M-theory.

Of course, because our 5-fold structures are nilpotents, squaring to zero, there is a sense in which the fifth term in each is redundant information, and so our 10 'dimensions' can be reduced to 8 overall, which is the number of fundamental units we started from. This would parallel the fact that, strings, with 1 dimension of space and 1 of time, or membranes, with 2 dimensions of space and 1 of time, exist in an 8-dimensional structure external to themselves; nilpotency then effectively removes the need for the structures by finding the conditions in which the extra dimensions are zeroed. A particularly interesting aspect of the two 5-fold reductions of the 8 fundamental units is the fact that they seem to correlate with two early modifications of Einstein's general relativity, which extended 4-dimensional space-time to 5 dimensions. The two separate theories of Kaluza and Klein have now been merged into one because they have the same mathematical structure, but originally they had different purposes, one being to explain the origin of mass and the other of electric charge.

Because of its circular compactification (it transforms a line length of ordinary space into a hosepipe shape with the extra dimension curling up in the hosepipe circumference), the fifth dimension introduces the $U(1)$ symmetry, which is the group symmetry which produces the characteristic Newton or Coulomb inverse-square law applied to both mass and electric charge. In our two 5-fold reductions to the nilpotent structure, we see that the 'extra' fifth dimensions of one is mass and the other is electric charge, just as in the Kaluza and Klein theories.

9.3 A little local difficulty

As it has become increasingly apparent that quantum mechanics requires some kind of superluminal communication between widely-separated particles and quantum systems, the strict defenders of the relativistic viewpoint have drawn a line in the sand, which reads quantum mechanical correlation does not violate relativity "because no information is transmitted". In the terms in which it is couched, this is true; but this is because the term 'information' has been commandeered to mean something very specific, which is quite different from what it means in more general terms. Information, in this context, means an interparticle interaction by means of boson transfer. The fastest that any particle, boson or fermion, real or virtual, can travel is the speed of light. In some sense all are doing exactly that; the speed of light is essentially the fundamental speed of particle motion. What makes most particles, other than a few bosons (photons and gluons) travel at *effectively* subluminal speed, or less than the speed of light, is the *zitterbewegung* which gives the particles their rest mass, which means that their travel includes motion in and out of vacuum.

'Information' in the more general sense, including that of information theory, includes knowledge about the state of a system from which one is separated, and in quantum mechanics, this is available *instantaneously* in a number of different ways. One of the most significant is provided by 'entangled states', which are now the basis of the emerging technology of quantum cryptography. Using this technology, we will soon be able to transmit information in the strict sense (*i.e.* as coded knowledge) faster than the speed of light. Though our *detection methods* may be limited by the speed of light, we can send information about the status of a measurement of spin over 140 kilometres and detect the result in a time faster than light could travel that distance; just one piece of coded information would be enough, for example, to tell us we had won the lottery. It doesn't make any sense, therefore, to maintain the fiction that 'information' refers only to bosonic transfer. It makes much more sense to suppose that there are two kinds of information: local information, which requires bosonic transfer and occurs no faster than the speed of light;

and nonlocal information, which is transmitted via a vacuum process and occurs, in principle, instantaneously. It is very likely, also, that, like many things in physics, they are dual processes.

If relativity is a property of bosonic transfer, then the transfer of nonlocal information by a vacuum process does not violate relativity. However, the long-held view that *all* information transfer and *all* physical processes occur at speeds limited by the speed of light becomes untenable. The reason why this view has so long prevailed is because naïve realism seems more readily acceptable to the human mind than a picture based on abstractions. The storybook picture of 'tangible' objects situated in a rigid structure called space (whatever that is) and interacting with each other by an equally tangible process has still not loosened its grip on our imagination, although not a single element of it makes sense in quantum mechanical terms. Even the idea that bosonic transfer is also a transfer of *energy* is by no means a certain fact; energy certainly *appears* to be transferred, but it may be that we are only looking at part of a process and that, if we could see the whole process, it would look very different. What *is* transferred is particle structure, in effect, discrete elements of charge of various kinds.

Now, if we have instantaneous transfer of nonlocal 'information', it seems reasonable to suppose that there must be a physical process involved. There must be something in the structures we have investigated that allows it. Perhaps we have been so obsessed with making a last stand for naïve realism that we have not even considered the one possibility that remains. Just one interaction has always failed to fit into the picture, yet in many ways it seems the obvious candidate for a nonlocal force. Mass-energy has nothing to do with discreteness; it is a continuous distribution throughout the entire universe, leaving no element of space untouched. If it undergoes local interactions, then it is difficult to see where the 'locality' is to be found. It sometimes seems to me that the theoretically-proposed Higgs field, which occupies every point in the universe with 246 GeV of energy, simply changes its structure as we define different places to be occupied or not with particles of various kinds. If every point of space has the same energy, then a 'transfer of energy' from point to point would be impossible in the strictest sense, though, since we never observe the entire energy at any point, we would observe changes of structure as though they were partial 'transfers of energy'.

If energy is *continuously* distributed across space, then it doesn't seem to make sense to have anything but an instantaneous connection between its units when acting as sources of a gravitational force. In what other way could we distinguish a continuous source from a discrete one? A continuous source suggests nonlocality as much as a discrete one requires the local. Also, absolute symmetry between the parameters suggests that mass must be distinguished from charge to the same extent that time is from space. A mathematical connection, as with time and space, suggests some formal similarities but disguises deep physical differences.

Gravity is extraordinarily weak compared with any of the other three forces. The gravitational force between two electrons at any distance is 10^{42} or a million trillion trillion trillion times weaker than the electric force between them. This is a colossal number, and the only reason why gravity figures as a major force on a universal scale is because it is everywhere and because it only has one sign of source, so unlike the electric force, where positive and negative charges largely cancel each other out, gravity adds up relentlessly in any system, and the bigger the system the bigger the gravity. In fact, factors of about 10^{40} were identified by Eddington and Dirac in the 1930s as relating quantities on cosmic and particle scales; essentially, the scale of gravity is naturally cosmic, while those of the other three forces are naturally particulate. Since the universe is the domain of the nonlocal, just as the point particle is the domain of the local, it makes sense that a force that is cosmic in scale and incredibly weak on a local scale should indeed be nonlocal.

One very curious fact is that the *cosmological constant*, a quantity which Einstein introduced into his general relativistic field equations in order to keep the universe from collapsing in on itself, has now been discovered experimentally as a small acceleration term to be applied to the cosmological redshift, the effect now generally taken to imply that the universe is expanding. In effect, the cosmological constant suggests a repulsive force or force-like effect, increasing with distance, making up about 68% of the universe's energy.[25] Such a cosmological constant could be predicted from quantum gravity. Although we have no successful theory of quantum gravity, we can still do an order of magnitude calculation of what we might expect it to be. Remarkably, it is nothing like the value found by the astronomers; some calculations suggest that it is a factor of 10^{120} too high; my own calculations raise this to 10^{123}. Now 10^{123} (which is a thousand followed by ten trillions) is a particularly interesting number. Seth Lloyd, using an information theory approach to the structure of the universe, calculates that this is the number of bit 'flips' (essentially quantum transitions or physical events) that could possibly take place during the lifetime of a universe structured like ours.[26] In effect, if a calculation is out by 10^{123}, it is not only wrong, but "not even wrong", in Pauli's famous words. It is as wrong as it could conceivably be.

Now, if a calculation produces a result like this, it isn't a problem, it's an opportunity. Something is telling us we must go in exactly the opposite direction to the one we had originally envisaged. We shouldn't be using quantum gravity at all, but should be going to something as different from this as it could be. Quantum calculations are what we do for *local* forces between isolated particles transmitted by boson exchange; perhaps we should be looking for something *nonlocal* between elements in a continuous distribution in which boson exchange has no part.

Previous experience should tell us that it is not nature's way that one description can apply to everything. We should expect opposites to occur in nature. We can only have the local because we also have the nonlocal, just as we can only have discreteness if we also have continuity. If we have local information, we should also have nonlocal information if we have local interactions, we should also have nonlocal ones. One of the ideas that string theory has been pursuing in recent years is gravity–gauge theory correspondence. Gauge theory here refers to the theory of the nongravitational interactions (electric, strong and weak). Gravity somehow supplies dual information to gauge theory. It is another example of the duality of the nonlocal and local. In fact, this is exactly what we would expect from our understanding of the continuous (gravitational) vacuum as being

partitioned into three distinct sections via the quaternion operators
k, **i**, **j** with this partitioning being associated with weak, strong and
electric interactions. It also follows from the fact that gravity is the
negative energy which cancels the positive energy of matter. Gravity,
in other words, is the vacuum force which negates real matter. This
is the gravity–gauge theory correspondence.

If it is true that gravity really is nonlocal, then we have a significant question to answer. How do we incorporate general relativity
and the effects that we know it describes correctly? The answer
must be that we should incorporate it in the same way that relativistic quantum mechanics incorporates special relativity, that is,
as an abstract mathematical structure which survives the context
in which it was first generated to become an integral part of a new
one. Here, the problem is that the space-time of quantum theory
and of observation and measurement cannot describe nonlocal interactions. Gravity affects this space-time, creating an effect which can
be described by equations of curvature. So gravity causes curvature
and this is described by the field equations. However, if gravity is
nonlocal, then curvature does not cause gravity. The positive feedback loop which causes so many problems for general relativity in
some interpretations is eliminated.

Of course, as we have seen already, nonlocal interactions can also
have local consequences. In the case of gravity, these are what we
describe as *inertia*, and what we actually observe in a gravitating
system is the local inertial reaction produced in discrete matter,
rather than gravity itself. Perhaps we can consider the gravity–inertia
connection as the result of making mass charge-like, just as wave–
particle duality is a result of a mathematical connection between
space and time that cannot be fully realised in physical terms. An
extended application of this idea, using a local coordinate system
which has been curved or rotated by gravity, and incorporating a
modified version of Mach's principle (the idea that the inertial properties of matter may be due to an interaction with the rest of the
matter in the universe), has been developed which predicts an effect
of the same kind as that now attributed to the cosmological constant, at approximately 67% of the energy of the universe; and it is

important to note that this was a *prediction*, predating the experimental discovery by a considerable period.[27]

One of the barriers to achieving a more unified picture of physics has been the failure of gravity to accommodate itself to quantum theory. The problem has been in the very fact that makes gravity most different from the other fundamental forces: the fact that for identical particles, gravity is an attractive force where all the others are repulsive. This may not seem particularly significant because the other forces can also be attractive if the interacting particles have opposite signs of charge. However, if we try to devise a quantum field theory of an attractive force between identical particles, we are obliged to use a spin 2 object as the mediating boson. For spin 1 bosons we can devise a renormalisable theory, that is, one in which the infinities can be eliminated by cancellation, but for spin 2 bosons we cannot. Though it has been claimed that M theory can solve the quantum gravity problem, this is more an expectation and a hope than a thing accomplished.

However, if gravity is really a nonlocal interaction and inertia its local and presumably quantised manifestation, there is a possibility that we may be able to devise a quantum theory which lends itself to renormalisation. This is because inertia is a repulsive force and therefore may be amenable to spin 1 boson representation. A discrete gravity theory by Manoelito de Souza and R. N. Silveira is based on a concept of *extended causality*, a development of the special relativistic 4-vector space-time including the proper time, which in our notation would be represented by a nilpotent $(ikt + i\mathbf{r} + j\tau)$.[28] The possibility of a nilpotent connection has made it especially interesting to us, and led to developments recorded in fuller detail elsewhere.[1]

According to de Souza and Silveira, a single object (particle or field) at two points in Minkowski space-time (represented by the 4-vector s) must satisfy the causality constraint $\Delta\tau^2 + \Delta s^2$, which in our interpretation becomes $\Delta(ikt + i\mathbf{r} + j\tau)^2 = 0$, where the symbol Δ denotes a change. "Extended causality" then applies when we shift τ and x by infinitesimal steps $d\tau$ and dx. Here, to make any sense of the idea, we need to add a little more technical detail than usual. Applying a massless scalar field will give us a discrete field equation, and a field source represented by a scalar charge to generate a 'graviton'-like object and a metric for a discrete gravitational field. In our terms, this could be an inertial field, with full quantisation supplied by the Dirac nilpotent, and the 'graviton'-like object being identified as a spin 1 boson or pseudo-boson, like the photon. Though this wouldn't be a quantised theory of gravity, it would supply a quantisation of the localised manifestation to which it relates, and may be the nearest approach we can make to applying quantum ideas to gravity.

Chapter 10

Conclusion

No one would start from where we are now in physics — two seriously incompatible theories seemingly on a collision course. Stephen Hawking is one of a number of physicists who say that we may never have a *unified* theory. But is this the kind of theory we really want? A union of the incompatibles? To put together unrelated ideas is just telling ourselves about our history. It doesn't indicate what the future should be. In some sense physics reached an impasse in 1973 with the Standard Model. Nothing discovered since has significantly changed our view of the subject. After a period of relatively rapid development we seem to have been unable to move forward.

Of course, it isn't obvious that being able to do physical science is anything to do with our evolution, and perhaps this is what Hawking means. In fact, we aren't particularly good at it. A small percentage of the human population can, by years of mental discipline and training in sustained thinking using abstract concepts, pick up skills and then proceed stumblingly to incrementally (and often by trial and error) extend the model. We can't alter the fact that the world happens to require mathematical description and this can often be extremely difficult. Even when we have mathematical equations, all things still have to be put through the human 'transducer' to make physical sense. However, there is no other species around that can do it at all, so we have no option but to try.

What we have called the storybook picture is one with different components arbitrarily put together in a composite. It cannot be a 'unified' theory. It is a compound one. Joining disparate things together is not unification. The minimum we need is a 'platform position', one that isn't necessarily the ultimate starting-point, but is a clear starting-point for everything else and has in itself a satisfying logical consistency, something like the chemists' periodic table or the biologists' genetic code and natural selection. We have this for portions of physics — conservation of energy and related principles for mechanics, Maxwell's equations for electromagnetic theory, the Dirac equation for relativistic quantum mechanics — but they don't connect into a coherent whole or in a simple way with the rest of physics. The physics of fundamental particles has a solid platform in the Standard Model, but it couldn't be said to have a satisfying logical consistency. After millions of successful confirmations by observation, the Standard Model is a safe bet, but it doesn't include gravity. It is also messy and far from elegant. It gets the right answers but doesn't seem to have any reason for its existence. This, of course, is an *opportunity*. The Standard Model contains many symmetries, generally treated as the result of highly sophisticated mathematics. However, the symmetries themselves are simple and look highly amenable to an entirely different approach.

What we should be looking for is a *foundational* theory, a theory from which everything in its disparity originates. Once we realise this, we can actually make massive progress on understanding space and time from a fundamental point of view, using the key idea of symmetry. Quantum mechanics gives us an opportunity to make ground on our understanding of fundamentals because it tells us something we didn't expect. It tells us something new, or at least indicates that there must be something new, and it makes sense to go in the direction in which it seems to be pointing.

The most important thing that quantum mechanics seems to be telling us is that the only true physics is abstract. This is entirely within the traditions of the historical development of the subject, though it is against our natural inclinations. We have to fully accept that quantum mechanics tells us that there is no 'tangible reality'. It is just an illusion. So it is completely wrong to try to construct a physics in which abstract ideas relate to tangible objects as though they could forever be maintained in separate spheres of knowledge — a world like this would mean we could never get to the beginning. It was what we had to do at first, but we need to take it now to the final stage. Reality includes concepts as well as 'things'. There is, in fact, no *tangible* real thing. Apparent tangibility is just the working out of abstract concepts. At the fundamental level, naïvely real tangible objects do not mix with abstract concepts. Matter, as far as we know, consists of points in space, not some extended solidity. Space isn't fixed either — there is no universal space in which everything moves. It is part of a package with the point singularities we consider as 'objects'.

It would be a natural thing to say that quantum mechanics makes no sense, whereas classical physics does. But 'natural' is not what physics is about. 'Natural' is about familiarity at a macroscopic level. We know we can learn nothing from this route. If physics is like quantum mechanics we have a chance. If physics is like classical physics then there is nothing we can do. We can't penetrate to any depth below the apparently real surface. So, we have to expect the unexpected, and look for an abstract solution that we wouldn't expect from naïve realism. The message of quantum mechanics is to go for the abstract in the most uncompromising manner. We have no hope of explaining quantum mechanics by going backwards to the pre-quantum position, and this in itself must push us in the right direction.

So what should we learn from quantum mechanics? First of all we should start thinking how strange it is that we accept as perfectly 'normal' that we should have some sort of 'material' objects in a semi-abstract concept we call 'space'. Obviously this is the position we have been left with by our slow development of physics starting from classical mechanics, and it has served us well for many centuries, but it can't be the final story. There is no way that we can advance to unification with such incompatible ideas.

The key way forward is to find the symmetries and dualities that are most significant in nature. We have found it in a symmetry between space, time, mass and charge, which turns out to be as exact as anything known in nature. Things that might have been difficulties turn out to be opportunities for penetrating even further into the fundamental structure.

We need space, time, mass and charge to be ideas of the same type, and we have four parameters, each of which has the same information as the other three combined, because it is a kind of mirror image. Time, mass and charge, for example, combine to produce an antispace. The same is true of mass, time and charge, and the combinations of the three parameters opposing each of them. The ultimate meaning seems to be that nature can't be characterised at the fundamental level. We are entitled to suppose that there isn't anything else outside of these parameters. With them as our guide, we can now establish that physics is an abstract subject very closely related to pure mathematics, in which things have the properties they do, not because we have to believe in a 'tangible' reality or a dominant observer or any model-dependent notion (such as complete discreteness), but because such properties are elements of an abstract structure that originates in symmetry. 'Tangibility' is part of our reality, not because it is more 'real' than abstract notions, but because it *as* real. Mathematics is not a 'tool' applied to physics for a strange reason unknown but an aspect of the extreme abstractness. The 'unreasonable' effectiveness of mathematics in physics and of physics in mathematics arise because both subjects emerge at the same abstract level.

The fundamental symmetry carries with it *its own* mathematics. The combination of two vector spaces (space and antispace or vacuum space) produces a very remarkable algebra, which is exactly what we need for relativistic quantum mechanics. Of the two spaces, the real one is observable, the vacuum one is not, but it carries the information concerning what happens in real space. We concentrate on 'real' space because that is how we observe and measure. It is also the most developed concept with the most structure. The spaces are commutative and dual because each contains the same information. The existence of two spaces is symptomatic of a doubling of the information, because the system and the universe need to negate each other. A manifestation of this effect is the ubiquity of the factor 2 in physical contexts.

Symmetry is important because it is the route to discovering absolutely nothing, as the entire structure of 'reality'. Zero totality is the most powerful of all universal constraints in physics. It is also the only possible starting-point, the only one which needs no further explanation, because explanation would be impossible. We have no idea what nothing actually is. It also has an infinite number of representations, which we shouldn't necessarily think of as occurring in a time sequence because time is part of the whole construct, not something in which the construct happens. When we say nothing, we mean it in the most abstract sense. Of course, this does not mean that nothing happens. Physicists are familiar with totality zero in cases where plenty of things actually happen — for example, in Newton's third law where the forces in a system are balanced, or when a gun being fired conserves the zero momentum before firing by creating nonzero amounts in opposite directions. Cosmologists frequently say that the universe came from 'nothing', but also that there were equal amounts of matter and antimatter in space at the beginning, which would mean starting with a finite amount of energy, as when an electron and antielectron pair emerges from two nonzero photons. Cosmologists also say that space was created at this moment, but if so, then so was antispace, which contains the antielectrons! The same is true for time. In fact, all events could be regarded as a 'creation' of this kind.

The 'vacuum' concept is more like a universal constraint, preserving zero totality, than something 'physical'. When the superposition of a fermion and vacuum adds to zero, it may seem strange to think of a 'real' particle and an 'unreal' abstract zero-complement, uniting to produce nothing, but the 'reality' of the particle, as we have seen, or, alternatively, the 'unreality' of the vacuum state is an illusion. The particle is as abstract as its zero-complement. Whatever happens, we can't expect to get *something* from nothing. Only nothing comes from nothing — *nihil ex nihilo fit*. If it starts at nothing that is where it will remain. The 'something' that we see all around us is only part of the picture from the *inside*. The total from the 'outside' that we can never completely comprehend is still nothing.

The actual structures lead us to an incredibly simple and powerful relativistic quantum mechanics defined only by an operator. In conventional quantum mechanics, the wavefunction is treated as a black box hiding behind a symbol ψ. Much of it is nonrelativistic, and necessarily distorted, but even the standard relativistic version uses a mathematical interpretation that is asymmetric and intrinsically meaningless. But a new, and automatically relativistic, quantum mechanics arises out of the symmetry of the parameters. Explanations of nonlocality and other quantum mechanical 'problems' are automatically included because these are properties of the parameters determined by the symmetry. Finding the nilpotent structure which provides the most efficient packaging of the separate information opens a Pandora's box. Not only does Schrödinger's cat escape from its dilemma (perhaps by tunnelling as we dither over making the measurement!), but we, as investigators, seem to obtain a more general release from many of the things which previously held us back from understanding more fundamental issues.

We need to be honest about exactly what theories tell us and cut away the inessential parts to get to the core. What is most significant is not necessarily the way things were originally done; there is sometimes unnecessary baggage which has to be left behind. For example, the two relativity theories are only compatible with each other and with quantum mechanics if we treat them as fundamentally mathematical theories on the same level as quantum mechanics, and are prepared to reject any physical interpretation of them which is incompatible with this and which has not been demonstrated by experiment. This is, in any case, exactly how they are structured. Classical relativity cannot be considered more fundamental than relativistic quantum mechanics or used to comment on its validity. We tend to learn about quantum mechanics and relativity separately, and then how Dirac put them together, and this comes over a little awkwardly, with a strange mathematical structure in the resulting equation, as though we hadn't properly combined the two. But this is looking at it the wrong way round. Essentially, nonrelativistic quantum mechanics and classical special relativity are both superseded by a theory which is more true than either separately, and is really a new theory, not a combination. In fact it is also easier to use than either original theory and makes more sense.

In the nilpotent theory, gravity, a product of the continuity of energy, is the means of keeping the universe holistically connected. It is the only force which is universal in every sense of the word, and in some respects is not really a force at all. Dealing with general relativity in this connection is probably less of an issue than is often assumed; it certainly doesn't make any sense to privilege it in understanding foundational subjects, as many have done without obvious success. When Einstein tried to develop a unification between gravity and electromagnetism, the forces seemed to be alike in many respects. It was natural to assume that any differences would soon be overcome. But complications arose very quickly with the discovery of two other forces, which behaved very differently from gravity and electromagnetism. It was like Hamilton trying to link real and imaginary numbers and finding that there were three types of the latter.

Even without this additional complication, things worked out differently to what had been expected, and differences between the sources of the original two forces emerged as well as similarities. Mass and charge are different in more respects than they are alike. This is also true of space and time. If we link space and time in one physical object, as many have thought must be possible, we lose sight of their differences. This 'physical' linking cannot happen in quantum mechanics, which rejects the notion of time as an observable, and this has often been thought of as the reason why the theory is incompatible with general relativity. As we have suggested, this problem can be overcome when we see how little the field equations of general relativity tell us about their physical meaning. This allows us to use them exactly in the way they were constructed, as mathematical equations about abstract concepts which do not necessarily have to imply the meanings that have been accrued over the years in the absence of testable solutions.

One of the key things about nilpotent quantum mechanics is that the nilpotent structure alone determines the nature of the fundamental interactions. Among many other results, it seems to make sense of particles as 'singularities' in its use of a double space, while the use of totality zero and the dual vacuum concept appears to supply the holistic aspect needed to explain many of the conundrums of quantum mechanics. Since nature often tries to reinvent itself at higher levels using formats that have been successful at lower ones, it seems likely that some of the structures may have relevance beyond the level at which quantum mechanics operates. Yes, we will start from here!

References

1. P. Rowlands, *Zero to Infinity: The Foundations of Physics* (World Scientific, 2007).
2. H. Minkowski, *Physikalische Zeitschrift*, **10**, 104–11 (1909), lecture on "Space and Time", Cologne, 21 September 1908, translated in H. A. Lorentz, A. Einstein, H. Minkowski and H. Weyl, *The Principle of Relativity* (Methuen and Company, Ltd., 1923), p. 104.
3. I. Newton, *Philosophia Naturalis Principia Mathematica* (London, 1687; 2nd edition, 1713; 3rd edition, 1726), Book Definitions, Scholium; translated by A. Motte as *Mathematical Principles of Natural Philosophy* (1729).
4. R. Penrose, in R. Penrose and C. J. Isham (eds.), *Quantum Concepts in Space and Time* (Clarendon Press, Oxford, 1986), p. 139.
5. J. D. Trimmer, The present situation in quantum mechanics: A translation of Schrödinger's "Cat Paradox paper", *Proceedings of the American Philosophical Society*, **124**, 323–38; also, J. A. Wheeler and W. H. Zurek (eds.), *Quantum Theory and Measurement* (Princeton University Press, New Jersey, 1983), Part I, Section I.11.
6. A. S. Eddington, *The Nature of the Physical World* (Macmillan, 1928).
7. W. Thomson, Lord Kelvin, *Baltimore Lectures*, 1884, reprinted in R. H. Kargon and P. Achinstein (eds.), *Kelvin's Baltimore Lectures and Modern Theoretical Physics* (MIT Press, 1987). W. Thomson, Lord Kelvin, *Baltimore Lectures on Molecular Dynamics and the Wave Theory of Light* (Cambridge University Press, Cambridge, 1904).
8. R. P. Graves, *The Life of Sir William Rowan Hamilton*, 3 vols. (Dublin, 1882–1891), Vol. 2, p. 435.
9. *ibid.*, Vol. 3, 239.
10. E. T. Bell, *Men of Mathematics* (Simon & Schuster, New York, 1937), paperback edition (1986).
11. M. J. Crowe, *A History of Vector Analysis: The Evolution of the Idea of a Vectorial System*, reprint edition (Dover Publication, New York, 1985), p. 42.
12. G. Santayana, *Reason in Common Sense*, Life of Reason, Vol. 1, (Charles Scribner's Sons, 1905–6), Chapter 12.
13. J. B. Kuipers, *Quaternions and Rotation Sequences* (Princeton University Press, 1998).

14. T. L. Hankins, *Sir William Rowan Hamilton* (Johns Hopkins University Press, 2004); C. C. Gillispie (ed.), *Dictionary of Scientific Biography* (Charles Scribner's Sons, 1975–1980).

15. P. G. Tait, *An Elementary Treatise on Quaternions*, 3rd edition (Cambridge University Press, 1890), p. vii, letter dated 12 April 1859.

16. G. J. Whitrow, *The Natural Philosophy of Time* (Nelson, London, 1961), pp. 135–57.

17. P. Coveney and R. Highfield, *The Arrow of Time* (W.H. Allen, London, 1990), pp. 28, 143, 144, 157.

18. A. Robinson, *Non-Standard Analysis*, revised edition (Princeton University Press, 1996).

19. R. Feynman, W. Morinigo and W. Wagner, *Feynman Lectures on Gravitation*, for academic year 1962–63, (Addison-Wesley Publishing Company, 1995), p. 10.

20. P. Atkins, *Creation Revisited* (Harmondsworth, 1994), p. 23.

21. P. A. M. Dirac, *The Principles of Quantum Mechanics*, 4th edition (Clarendon Press, Oxford, 1958).

22. G. 't Hooft, Dimensional reduction in quantum gravity, in A. Ali, J. Ellis and S. Randjbar-Daemi (eds.), *Highlights of Particle and Condensed Matter Physics: Salamfest,* Proceedings of the Conference, ICTP, Trieste, Italy, 8–12 March 1993, World Scientific Series in 20th Century Physics, Vol. 4 (World Scientific, Singapore, 1993), pp. 284–96; L. Susskind, *J. Math. Phys.*, **36**, 637 (1995).

23. R. P. Feynman, The development of the space-time view of quantum electrodynamics, in *Nobel Lectures: Physics 1963–70* (Elsevier Publishing Company, Amsterdam, 1972).

24. A. Faroggi, Lectures, University of Liverpool.

25. S. Lloyd, Computational capacity of the universe, *Phys. Rev. Lett.*, **88** (23), 237901 (2002); V. Giovannetti, S. Lloyd and L. Maccone, Quantum-enhanced measurements: Beating the standard quantum limit, *Science*, **306**, 1330–6, 2004.

26. P. A. R. Ade *et al.* (Planck Collaboration), *A&A* (2013), doi:10.1051/0004-6361/201321529, arXiv:1303.5062; P. A. R. Ade *et al.* (Planck Collaboration), *A&A* (2013), doi:10.1051/0004-6361/201321591, arxiv:1303.5076.

27. P. Rowlands, *A Revolution Too Far* (PD Publications, Liverpool, 1994); and several earlier works.

28. M. de Souza and R. N. Silveira, Discrete and finite general relativity, *Class. Quantum Grav.*, **16**, 619 (1999).

Index